Collins

AQA GCSE (9-1)

Physics

Grade 5 Booster Workbook

Stephanie Grant
Lynn Pharaoh

William Collins' dream of knowledge for all began with the publication of his first book in 1819. A self-educated mill worker, he not only enriched millions of lives, but also founded a flourishing publishing house. Today, staying true to this spirit, Collins books are packed with inspiration, innovation and practical expertise. They place you at the centre of a world of possibility and give you exactly what you need to explore it.

Collins. Freedom to teach

HarperCollins*Publishers*
The News Building
1 London Bridge Street
London SE1 9GF

HarperCollins*Publishers*
Macken House, 39/40 Mayor Street Upper,
Dublin 1, D01 C9W8,
Ireland

Browse the complete Collins catalogue at www.collins.co.uk

First edition 2016

10 9 8 7

MIX
Paper
FSC www.fsc.org FSC™ C007454

ISBN 978-0-00-819438-3

www.collins.co.uk

A catalogue record for this book is available from the British Library

Commissioned by Joanna Ramsay
Project managed by Sarah Thomas and Siobhan Brown
Development edited by Gillian Lindsey
Copy edited by Gwynneth Drabble
Proofread by David Hemsley
Typeset by Jouve India Pvt Ltd
Artwork by Jouve India Pvt Ltd
Cover design by We are Laura and Jouve
Cover image: Shutterstock/Mihai Simonia
Printed by Ashford Colour Ltd.

Contents

Introduction

This workbook will help you build your confidence in answering Physics questions for GCSE Physics and GCSE Combined Science.

It gives you practice in using key scientific words, writing longer answers, answering synoptic questions as well as applying knowledge and analysing information.

The questions also cover required practicals, maths skills and synoptic questions – look out for the tags which will help you to identify these questions.

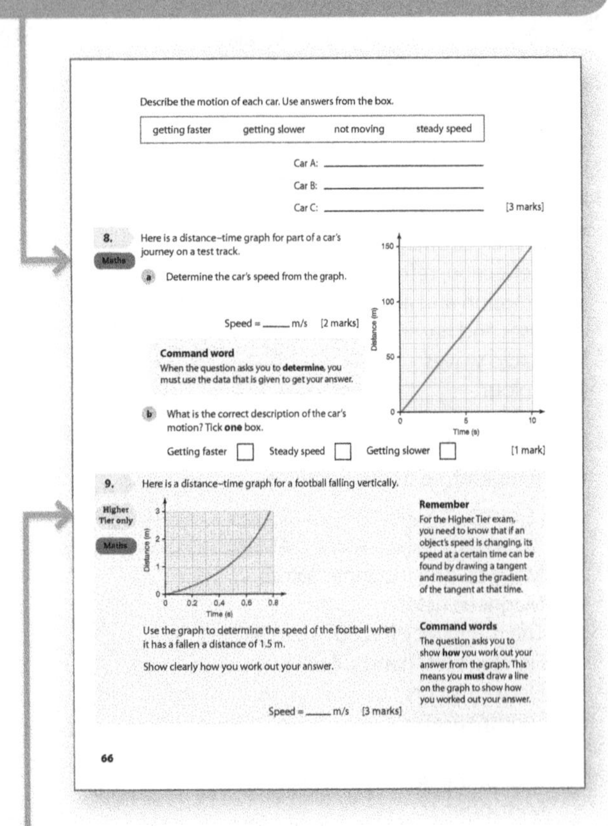

Describe the motion of each car. Use answers from the box.

getting faster	getting slower	not moving	steady speed

Car A: ____

Car B: ____

Car C: ____ [3 marks]

8. Maths

Here is a distance–time graph for part of a car's journey on a test track.

a Determine the car's speed from the graph.

Speed = ____ m/s [2 marks]

Command word
When the question asks you to **determine**, you must use the data that is given to get your answer.

b What is the correct description of the car's motion? Tick **one** box.

Getting faster ☐ Steady speed ☐ Getting slower ☐ [1 mark]

9. Higher Tier only Maths

Here is a distance–time graph for a football falling vertically.

Remember
For the Higher Tier exam, you need to know that if an object's speed is changing, its speed at a certain time can be found by drawing a tangent and measuring the gradient of the tangent at that time.

Use the graph to determine the speed of the football when it has fallen a distance of 1.5 m.

Show clearly how you work out your answer.

Speed = ____ m/s [3 marks]

Command words
The question asks you to show **how** you work out your answer from the graph. This means you **must** draw a line on the graph to show how you worked out your answer.

66

You will find all the different question types in the workbook so you can get plenty of practice in providing short and long answers.

Higher Tier content is clearly marked throughout.

Learn how to answer test questions with annotated worked examples.

This will help you develop the skills you need to answer questions.

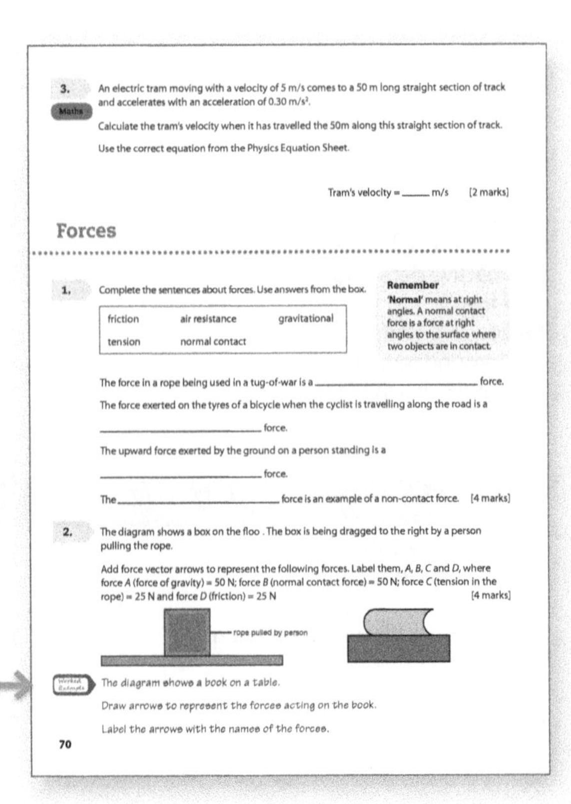

3. Maths

An electric tram moving with a velocity of 5 m/s comes to a 50 m long straight section of track and accelerates with an acceleration of 0.30 m/s^2.

Calculate the tram's velocity when it has travelled the 50m along this straight section of track.

Use the correct equation from the Physics Equation Sheet.

Tram's velocity = ____ m/s [2 marks]

Forces

1. Complete the sentences about forces. Use answers from the box.

friction	air resistance	gravitational
tension	normal contact	

Remember
'Normal' means at right angles. A normal contact force is a force at right angles to the surface where two objects are in contact.

The force in a rope being used in a tug-of-war is a ____ force.

The force exerted on the tyres of a bicycle when the cyclist is travelling along the road is a ____ force.

The upward force exerted by the ground on a person standing is a ____ force.

The ____ force is an example of a non-contact force. [4 marks]

2. The diagram shows a box on the floo . The box is being dragged to the right by a person pulling the rope.

Add force vector arrows to represent the following forces. Label them, A, B, C and D, where force A (force of gravity) = 50 N; force B (normal contact force) = 50 N; force C (tension in the rope) = 25 N and force D (friction) = 25 N [4 marks]

Worked Example

The diagram shows a book on a table.

Draw arrows to represent the forces acting on the book.

Label the arrows with the names of the forces.

70

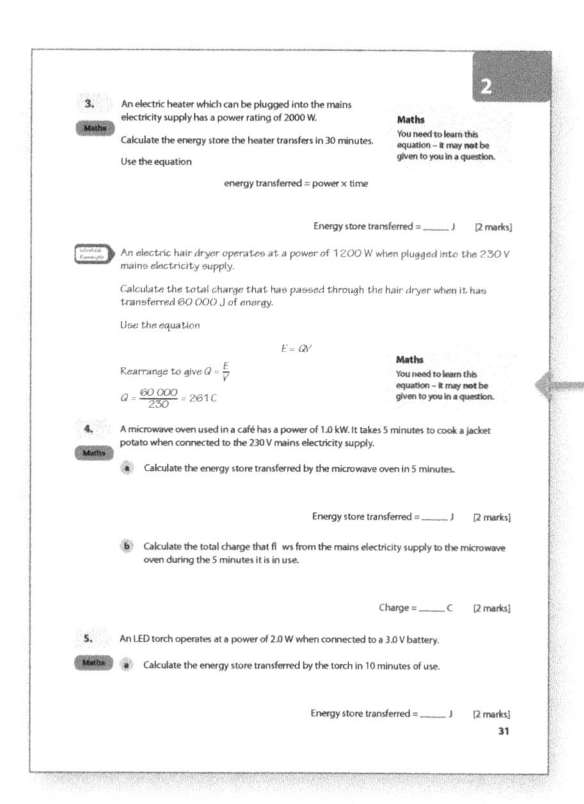

2

3. An electric heater which can be plugged into the mains electricity supply has a power rating of 2000 W.

Maths

Calculate the energy store the heater transfers in 30 minutes.

Use the equation

energy transferred = power × time

Maths
You need to learn this equation – it may **not** be given to you in a question.

Energy store transferred = ______ J [2 marks]

Worked Example

An electric hair dryer operates at a power of 1200 W when plugged into the 230 V mains electricity supply.

Calculate the total charge that has passed through the hair dryer when it has transferred 60 000 J of energy.

Use the equation

$E = QV$

Rearrange to give $Q = \frac{E}{V}$

$Q = \frac{60\,000}{230} = 261\text{ C}$

Maths
You need to learn this equation – it may **not** be given to you in a question.

4. A microwave oven used in a café has a power of 1.0 kW. It takes 5 minutes to cook a jacket potato when connected to the 230 V mains electricity supply.

Maths

a) Calculate the energy store transferred by the microwave oven in 5 minutes.

Energy store transferred = ______ J [2 marks]

b) Calculate the total charge that fl ws from the mains electricity supply to the microwave oven during the 5 minutes it is in use.

Charge = ______ C [2 marks]

5. An LED torch operates at a power of 2.0 W when connected to a 3.0 V battery.

Maths

a) Calculate the energy store transferred by the torch in 10 minutes of use.

Energy store transferred = ______ J [2 marks]

31

There are lots of hints and tips to help you out. Look out for tips on how to decode command words, key tips for required practicals and maths skills, and common misconceptions.

The amount of support gradually decreases throughout the workbook. As you build your skills you should be able to complete more of the questions yourself.

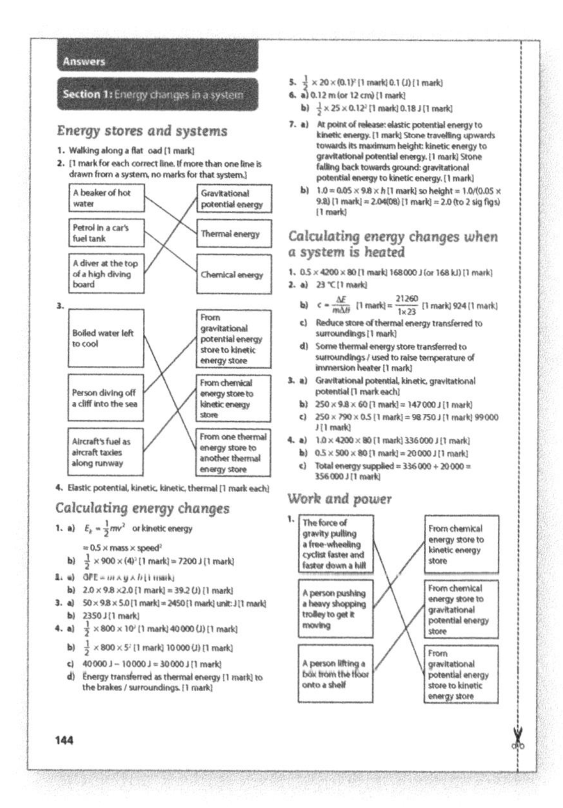

Answers

Section 1: Energy changes in a system

Energy stores and systems

1. Walking along a flat oad [1 mark]
2. [1 mark for each correct line. If more than one line is drawn from a system, no marks for that system.]

3.

4. Elastic potential, kinetic, kinetic, thermal [1 mark each]

Calculating energy changes

1. a) $E_k = \frac{1}{2}mv^2$ or kinetic energy $= 0.5 \times \text{mass} \times \text{speed}^2$
   b) $\frac{1}{2} \times 900 \times (4)^2$ [1 mark] = 7200 J [1 mark]
2. a) GPE = $m \times g \times h$ [1 mark]
   b) $2.0 \times 9.8 \times 2.0$ [1 mark] = 39.2 (J) [1 mark]
3. a) $50 \times 9.8 \times 5.0$ [1 mark] = 2450 [1 mark] unit: J [1 mark]
   b) 2350 J [1 mark]
4. a) $\frac{1}{2} \times 800 \times 10^2$ [1 mark] 40 000 (J) [1 mark]
   b) $\frac{1}{2} \times 800 \times 5^2$ [1 mark] 10 000 (J) [1 mark]
   c) 40 000 J – 10 000 J = 30 000 J [1 mark]
   d) Energy transferred as thermal energy [1 mark] to the brakes / surroundings. [1 mark]

5. $\frac{1}{2} \times 20 \times (0.1)^2$ [1 mark] 0.1 (J) [1 mark]
6. a) 0.12 m (or 12 cm) [1 mark]
   b) $\frac{1}{2} \times 25 \times 0.12^2$ [1 mark] 0.18 J [1 mark]
7. a) At point of release: elastic potential energy to kinetic energy. [1 mark] Stone travelling upwards towards its maximum height: kinetic energy to gravitational potential energy. [1 mark] Stone falling back towards ground: gravitational potential energy to kinetic energy. [1 mark]
   b) $1.0 = 0.05 \times 9.8 \times h$ [1 mark] so height = 1.0/(0.05 × 9.8) [1 mark] = 2.04(08) [1 mark] = 2.0 (to 2 sig figs) [1 mark]

Calculating energy changes when a system is heated

1. $0.5 \times 4200 \times 80$ [1 mark] 168 000 J (or 168 kJ) [1 mark]
2. a) 23 °C [1 mark]
   b) $c = \frac{\Delta E}{m\Delta\theta}$ [1 mark] $= \frac{21260}{1 \times 23}$ [1 mark] 924 [1 mark]
   c) Reduce store of thermal energy transferred to surroundings [1 mark]
   d) Some thermal energy store transferred to surroundings / used to raise temperature of immersion heater [1 mark]
3. a) Gravitational potential, kinetic, gravitational potential [1 mark each]
   b) $250 \times 9.8 \times 60$ [1 mark] = 147 000 J [1 mark]
   c) $250 \times 790 \times 0.5$ [1 mark] = 98 750 J [1 mark] 99 000 J [1 mark]
4. a) $1.0 \times 4200 \times 80$ [1 mark] 336 000 J [1 mark]
   b) $0.5 \times 500 \times 80$ [1 mark] = 20 000 J [1 mark]
   c) Total energy supplied = 336 000 + 20 000 = 356 000 J [1 mark]

Work and power

1.

144

There are answers to all the questions at the back of the book. You can check your answers yourself or your teacher might tear them out and give them to you later to mark your work.

Energy stores and systems

1. Which movement does **not** change a person's store of gravitational potential energy? Tick **one** box.

Riding a bicycle up a hill ☐

Running down the stairs ☐

Walking along a flat road ☐

Using a lift to change floor in a building ☐

[1 mark]

2. Draw **one** line from each system to its store of energy.

System	Store
A beaker of hot water	Gravitational potential energy
Petrol in a car's fuel tank	Thermal energy
A diver at the top of a high diving board	Chemical energy

[3 marks]

3. Draw **one** line from each system to the change in that system's energy store as time passes.

System	Change in energy stored
Boiled water left to cool	From gravitational potential energy store to kinetic energy store
Person diving off a cliff into the sea	From chemical energy store to kinetic energy store
Aircraft's fuel as aircraft taxies along runway	From one thermal energy store to another thermal energy store

[3 marks]

4. Complete the sentences about how the different energy stores change as a stone is fired at a wall from a catapult. Use words from the box.

thermal	kinetic	elastic potential

As the catapult is stretched, its store of ______________________

energy is increased. As the catapult is fired this store of energy is transferred to the

______________________ energy store of the stone. As the stone hits the wall,

the ______________________ energy store of the stone decreases and the

______________________ energy store of the wall increases.

[4 marks]

Calculating energy changes

1. Maths

a Write down the equation which links speed, kinetic energy and mass.

_______________________________ [1 mark]

Maths

You need to learn both these equations – they may **not** be given to you in a question.

b A car of mass 900 kg is travelling at 4 m/s.

Calculate the amount of kinetic energy it stores.

Kinetic energy store = _______ J [2 marks]

2. Maths

a Write down the equation which links gravitational field strength, gravitational potential energy, height and mass.

_______________________________ [1 mark]

b A 2.0 kg box is lifted onto a 2.0 m high shelf. By how much does its gravitational potential energy change?

Gravitational field strength = 9.8 N/kg.

Change in store of gravitational potential energy = _______ J [2 marks]

3. Maths

a A diver of mass 50 kg stands on a 5.0 m high diving board.

Calculate the change in the diver's gravitational potential energy as she falls from the board into the water.

Give the unit for your answer.

Gravitational field strength = 9.8 N/kg.

Change in store of gravitational potential energy = _______

Unit _______ [2 marks]

b As the diver falls towards the swimming pool below, energy is transferred from a store of gravitational potential energy to a store of kinetic energy and the thermal energy store of the surrounding air.

If 100 J of her gravitational potential energy store is transferred to the thermal energy store of the air during the dive, how much kinetic energy does she have as she enters the water?

Kinetic energy store = ______ J [1 mark]

4. Maths

a A car of mass 800 kg is travelling at 10 m/s. The driver applies the car's brakes, reducing the car's speed to 5 m/s.

Use the equation $E_k = \frac{1}{2}mv^2$ to calculate the car's initial store of kinetic energy.

Initial kinetic energy store = ______ J [2 marks]

b Calculate the car's kinetic energy store when its speed has been reduced to 5 m/s.

Kinetic energy store = ______ J [2 marks]

c Calculate the change in the car's kinetic energy store during the time the brakes are applied.

Change in store of kinetic energy = ______ J [1 mark]

d Describe what happened to the kinetic energy store lost in part c.

______

______ [2 marks]

5. Maths

A spring has a spring constant of 20 N/m. It is extended by 0.1 m.

By how much does its elastic potential energy change?

Use the equation:

$E_e = \frac{1}{2}ke^2$.

Change in store of elastic potential energy = ______ J [2 marks]

Remember

Although you are given the equation for elastic potential energy here, it is useful to understand what is meant by 'extension' and 'spring constant'. When a spring is stretched, the **increase** in its length is called the **extension**. A spring with a large value for '*k*' is much harder to stretch than a spring with a small value for '*k*'.

6. **a** The length of an unstretched spring is 0.020 m. The spring has a spring constant of 25 N/m. The spring is stretched so that its length is 0.140 m.

Calculate the spring's extension.

Extension = ______ m [1 mark]

b Calculate the elastic potential energy stored in the stretched spring.

Elastic potential energy store = ______ J [2 marks]

7. **a** A stone is fired vertically upwards using a catapult.

Maths

Describe the energy transfers that take place at the following points in the stone's path.

At the point of release:

____

The stone is travelling upwards towards its maximum height:

____

The stone is falling back towards the ground:

____ [3 marks]

b The energy stored in the stretched catapult is 1.0 J. The mass of the stone in the catapult is 50 g. If all the elastic energy in the elastic potential energy store is transferred to the stone's gravitational potential energy store, calculate the height above the point of release the stone will reach.

Give your answer to two significant figures.

Gravitational field strength = 9.8 N/kg.

Maths

To **round** a number to two significant figures, look at the **third** digit. Round up if the digit is 5 or more, and round down if the digit is 4 or less.

For example, 1.62 rounds to 1.6.

Height = ______ m [4 marks]

Calculating energy changes when a system is heated

1. Water has a specific heat capacity of 4200 J/kg °C. How much energy store is transferred to 500 g of water when its temperature increases from 20.0 °C to 100.0 °C?

Use the correct equation from the Physics Equation Sheet.

Maths

This question gives the mass in grams, so you first need to change the mass into kilograms.

Energy store transferred = ________________ J [2 marks]

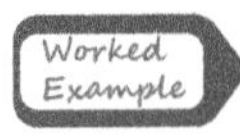

How much energy is needed to increase the temperature of 2.0 kg of water from 20 °C to 70 °C?

The specific heat capacity of water is 4200 J/kg °C.

The temperature change of the water, $\Delta q = 70° - 20° = 50$ °C

Energy needed = $mc\Delta q$

$= 2.0 \times 4200 \times 50 = 420\ 000$ J

2. Practical

A student uses the apparatus shown in the diagram to measure the specific heat capacity of aluminium. A heater slots into the larger hole in the aluminium block and the thermometer slots in the smaller hole. The aluminium block is wrapped in expanded polystyrene, which is an insulator.

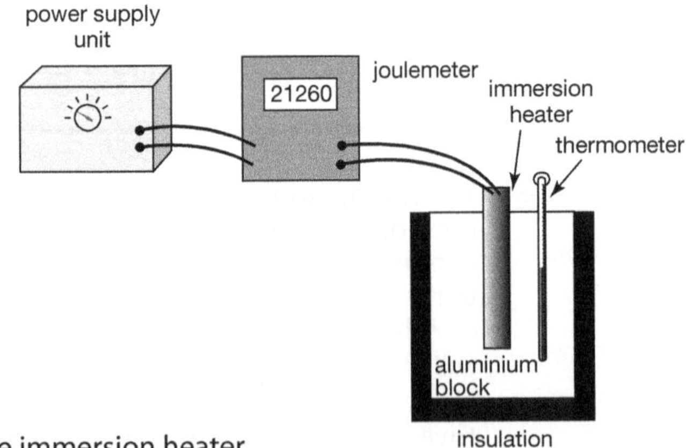

The joulemeter records the amount of energy transferred by the electric current to the thermal energy store of the immersion heater.

The student switches the power supply unit on for 10 minutes. The table shows the results.

Mass of block (kg)	Joulemeter reading (J)	Initial temperature of block (°C)	Temperature of block after heating (°C)
1.00	21 260	18.0	41.0

a Calculate the temperature rise of the block.

Temperature rise = ________ °C [1 mark]

b Use the data in the table to calculate the specific heat capacity, c, of aluminium.

Use the correct equation from the Physics Equation Sheet.

Give your answer to three significant figures.

Maths

You will need to **rearrange** the equation for this calculation. This means putting specific heat capacity on its own, on one side of the equals sign.

The equation is in the form $a = b \times c \times d$. To find c, divide both sides of the equation by b. Then divide both sides of the equation by d.

Specific heat capacity = ______ J/kg °C [3 marks]

c Describe the purpose of the insulation wrapped around the aluminium block.

____

____ [1 mark]

d The accepted value for the specific heat capacity of aluminium is 910 J/kg °C.

Suggest a reason why the value obtained from the student's data is greater than the accepted value.

____

____ [1 mark]

3. Maths

a A granite boulder of mass 250 kg falls from a cliff 60 m high to the ground below.

Complete the sentences to describe the energy transfers. Use words from the box.

thermal	gravitational potential	kinetic

At the top of the cliff, the boulder has a store of

______________________ energy. As the boulder falls

its store of ______________________ energy increases and its store of

______________________ decreases. [3 marks]

b Calculate the change in the boulder's gravitational potential energy as it falls from the top of the cliff to the ground.

Gravitational field strength = 9.8 N/kg.

Change in store of gravitational potential energy = ______________ J [2 marks]

c On striking the ground, the energy in the kinetic energy store of the boulder is transferred to the store of thermal energy of the boulder and the surrounding air. The temperature of the boulder increases by 0.5 °C.

Use the equation

change in thermal energy = mass × specific heat capacity × temperature change

to calculate the increase in the boulder's thermal energy store. Give your answer to two significant figures. The specific heat capacity of granite is 790 J/kg °C.

Increase in store of thermal energy = ____________ J [3 marks]

4. Maths

a An electric kettle contains 1.0 kg of water at 20 °C. The kettle is switched on until the water reaches 100 °C. Calculate the increase in the thermal energy store of the water.

The specific heat capacity of water is 4200 J/kg °C.

Maths

In an integer, zero is counted as a significant figure **only** if it is between two other non-zero significant digits. For example, the zeros in 10 000 are **not** significant.

Increase in thermal energy store of the water = ____________ J [2 marks]

b The kettle is made from stainless steel and has a mass of 0.50 kg. The temperature of the kettle is also increased from 20 °C to 100 °C. Calculate the increase in the thermal energy store of the kettle.

The specific heat capacity of stainless steel is 500 J/kg °C.

Increase in thermal energy of kettle = ____________ J [2 marks]

c Calculate the total energy store that would have to be supplied to the kettle from the electricity supply.

Total energy store supplied = ____________ J [1 mark]

Work and power

1. Work is done when a force moves an object from one place to another. When work is done, energy is transferred.

Draw **one** line to match each example of work to the energy transfer that takes place.

Maths
You need to learn the equations for work done and power – they may **not** be given to you in a question.

Example of work	Energy transfer
The force of gravity pulling a free-wheeling cyclist faster and faster down a hill	From chemical energy store to kinetic energy store
A person pushing a heavy shopping trolley to get it moving	From chemical energy store to gravitational potential energy store
A person lifting a box from the floor onto a shelf	From gravitational potential energy store to kinetic energy store

[3 marks]

2. Synoptic

a A child pulling a sledge along a horizontal path exerts a force of 10 N for 10 s over a distance of 15 m.

Use the equation

work done = force × distance

to calculate the work he does in pulling the sledge.

Synoptic
In the final exam, some of the marks will be for **connecting** your knowledge from different areas of physics. To answer this question you need to **link** knowledge of energy store transfer with ideas about forces (section 3).

Work done = ______ J [2 marks]

b Use the equation

$$\text{power} = \frac{\text{work done}}{\text{time}}$$

to calculate the child's useful output power in moving the sledge.

Useful output power = ______ W [2 marks]

3. **Synoptic**

a When a driver of a car applies the brakes, a force of 4000 N brings the car to a stop over a distance of 50 m.

Calculate the work done by the braking force.

Work done = ____________ J [2 marks]

b Calculate the braking power if the time taken to stop is 5.0 s.

Give your answer in kilowatts.

Braking power = ______ kW [2 marks]

4. **Maths**

a An electric motor-driven crane lifts an 800 kg load a distance of 10 m. Calculate the gain in gravitational potential energy of the load.

Gravitational field strength = 9.8 N/kg.

Energy = ____________ J [2 marks]

b Calculate the useful power output of the crane if it takes 20 s to raise the load.

Useful power output = ____________ W [2 marks]

Conservation of energy

1. What name is given to the transfer of energy to a store from which it cannot be recovered? Tick **one** box.

Conduction ☐ Cooling ☐ Dissipation ☐ [1 mark]

2. Which one of the statements about energy is **incorrect**? Tick **one** box.

Energy can be usefully transferred from one energy store to another. ☐

Energy can be stored for future use. ☐

Energy can be destroyed. ☐

Energy can be dissipated. ☐ [1 mark]

3. A cyclist is travelling at a steady speed of 9 m/s along a flat road. The forces of friction and air resistance transfer 165 J of energy store away from the cyclist to the surroundings every second.

How much useful output energy must the cyclist generate every second to maintain her steady speed?

Useful output energy generated every second = ______ J [1 mark]

4. A child is travelling at a steady speed along a horizontal path with a push scooter. When she stops pushing, the scooter gradually slows down and stops.

Complete the sentences to explain what happens to the kinetic energy store of the scooter as it slows down.

When the child stops pushing the kinetic energy store of the scooter

______________________________.

Energy is transferred away from the scooter because work is done by

______________________________.

The total energy store of the system (the child, the scooter and the surroundings) is

______________________________.

This means the energy in the kinetic energy store of the scooter has been transferred to

______________________________. [4 marks]

Ways of reducing unwanted energy transfers

1. Friction in the bearings of a bicycle wheel dissipates energy to the surroundings.

How can this unwanted energy transfer can be reduced?

______________________________ [1 mark]

2. **Practical** A group of students used the apparatus in the diagram to investigate how effective different materials are as thermal insulators. They also have a stopclock.

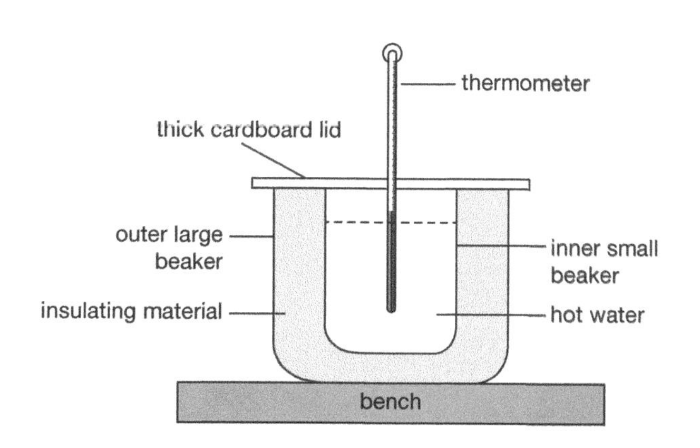

a Describe a method the students could use to investigate how effective different materials are as thermal insulators.

Describe how you would do the investigation and the measurements you would make.

Describe how you would make it a fair test.

Command term

This is an extended writing question. You could use bullet points or subheadings but your points **must** be in full sentences and in an order that makes sense.

____

____

____

____

____

____

____

____

____

____ [6 marks]

b The results for two of the materials are shown in the graph.

Use the graph to decide which of these two materials is the better thermal insulator.

Give a reason for your answer.

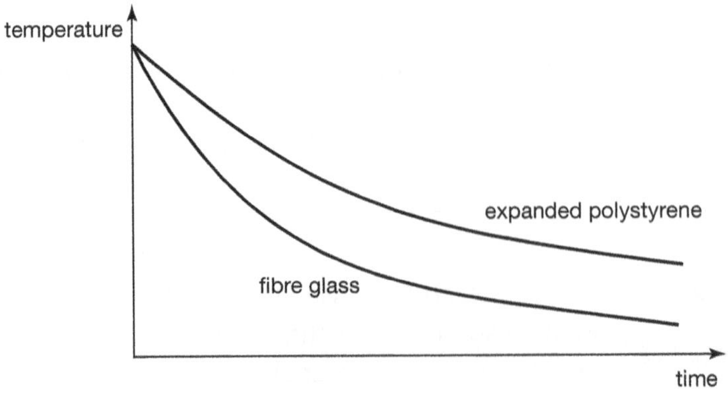

temperature

expanded polystyrene

fibre glass

time

____

____ [2 marks]

Efficiency

1. Maths

a Complete the sentences about energy changes in a car's diesel engine. Use words from the box.

thermal	kinetic	chemical

The diesel fuel in a car's fuel tank is a store of ______________ energy. The car's engine transfers energy from the store in the fuel to the ______________ energy store of the car. However some of the energy is dissipated to the surroundings, which increases the ______________ energy store of the surroundings. [3 marks]

b When the car's tank is full of diesel fuel it stores 1000 MJ of energy. When all the fuel has been used, the car's engine has transferred 420 MJ of this energy to the car to increase its store of kinetic energy. How much energy is dissipated to the surroundings?

Energy dissipated = ______ MJ [1 mark]

c Use the equation: $\text{efficiency} = \dfrac{\text{useful output energy transfer}}{\text{total input energy transfer}}$

to calculate the efficiency of the car's engine.

Give your answer as both a decimal and as a percentage.

Maths
You need to learn the equations for efficiency – they may **not** be given to you in a question.

Efficiency as a decimal = ______

Efficiency as a percentage = ______ % [2 marks]

2. Maths

A hybrid car has an efficiency of 80%.

Calculate the useful output energy if the engine is supplied with 500 MJ.

Maths
Efficiency values are often given as percentages. It is useful to remember that percentages are fractions of 100. So 40% is the fraction $\frac{40}{100}$.

Useful output energy = ______ MJ [3 marks]

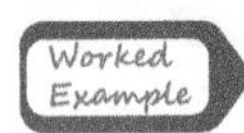

The efficiency of car's diesel engine is 40%. Calculate the useful output energy if the total input energy of a tank of fuel is 1500 MJ.

40% as a fraction is $\frac{40}{100}$.

Substituting into the efficiency equation gives:

$$\frac{40}{100} = \frac{\text{useful output energy transfer}}{1500}.$$

So the useful output energy $= \frac{40}{100} \times 1500$

$= 600$ MJ

3. Maths

A gas-fired power station has an efficiency of 50%. How much input energy store must be transferred from the gas for every 1000 MJ of useful output electrical energy?

Input energy store = ____________ MJ [3 marks]

Maths

You will need to rearrange the equation to find the useful output energy. This means putting useful output energy on its own, on one side of the equals sign.

If an equation is given in the form $a = b/c$, to find b, first multiply both sides by c.

Maths

You could also work out the answer to this question using ratios. If the efficiency is 50%, only half of the input energy is usefully transferred so the input energy transfer must be twice as much as the useful output energy transfer.

Whichever method you use, always show how you work it out. Then you may still get some marks even if your final answer is wrong.

4. Higher Tier only

The efficiency of a house heating system can be improved if less thermal energy is transferred from the house to the surrounding air.

Give **two** ways that this can be achieved.

____

____ [2 marks]

National and global energy resources

1. The percentage of the total energy used in generating electricity in the UK in 2013, 2014 and 2015 for the main four energy resources is shown in this table .

Energy resource	2013 total consumption (%)	2014 total consumption (%)	2015 total consumption (%)
coal	36	29.1	20.5
gas	27	30.2	30.2
nuclear	20	19.0	21.5
renewable	14.9	19.2	25.3

a Which **two** of the fuels in this table can be described as fossil fuels?

______________________________ [1 mark]

b Name a substance used as fuel in a nuclear power station.

______________________________ [1 mark]

c Coal, gas and nuclear are described as non-renewable energy resources.

Explain what is meant by a 'non-renewable energy resource'.

____

______________________________ [2 marks]

d Describe the trends in the data shown in the table.

____

____

____

______________________________ [3 marks]

Literacy

A **trend** is a pattern in data. Don't just say 'it goes up' or 'it goes down'. **Compare** the increase or decrease using data from the table.

e Compare the environmental effects of using coal or nuclear fuel for generating electricity.

____

____

____

_______________ [3 marks]

Command words

A **compare** question means give similarities and differences. You will need to mention both parts in the comparison. Good words to use in your answers are 'both', 'however', 'whereas' and 'but'. Remember, 3 marks means 3 points are needed.

2. In 2014, renewable energy sources provided 19.2% of the energy used for generating electricity in the UK. The four main types of renewable energy resource used in electricity generation in 2014 were onshore wind, offshore wind, solar panels (photovoltaic cells) and bio-fuel.

a Give an example of a bio-fuel that can be used in the generation of electricity.

_______________ [1 mark]

b The reliability of power stations is measured by their ability to maintain a constant output of electricity.

Give **one** reason why it is important that the energy resources used to generate electricity are reliable.

____

_______________ [1 mark]

c Describe the **problems** of reliability of wind power and solar (photovoltaic) panels as energy resources for generating electricity.

Wind power: _______________

Solar power: _______________ [2 marks]

3. The graph compares the uses of energy in the UK by transport, homes and industry from 1970 to 2014. In the same period, the UK population increased from 55.7 million in 1970 to 64.6 million in 2014.

Compare the trends shown in this graph.

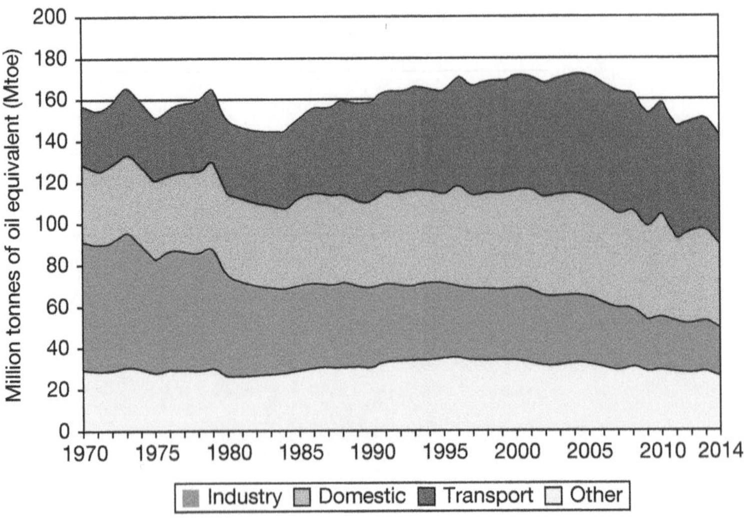

Literacy

This is an extended writing question. To get full marks you need to link your sentences together in a sensible order.

When the question asks you to **suggest**, you must link what you already know to the new information you are given.

Literacy

Remember, a trend is a **pattern** in data. You need to **compare** the trends, which means give similarities **and** differences. For example, if two categories both increase, which shows the largest increase? You also need to **link** your sentences together in a sensible order to get full marks.

Suggest possible explanations.

____

____

____

____

____

____

____ [6 marks]

4. The UK uses less energy today than in 1970 even though the UK's population has increased by nearly 10 million. In the last 10 years, global energy consumption per person has increased by approximately 10%.

Suggest **two** reasons why the global trend in energy consumption is very different to that in the UK.

____

____ [2 marks]

Circuit diagrams

1. Draw **one** line from each circuit symbol to the correct description.

Symbol	Description
(cell symbol)	open switch
(open switch symbol)	cell
(lamp symbol)	resistor
(resistor symbol)	lamp

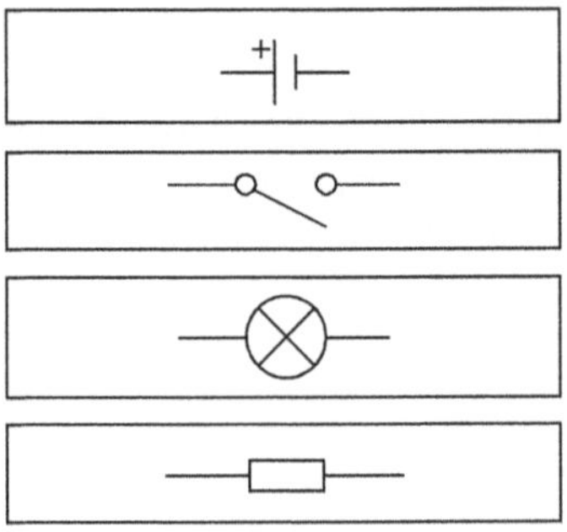

[4 marks]

2. A student is asked to build the circuit in the diagram. The components in the circuit are joined together using connecting leads. [1 mark]

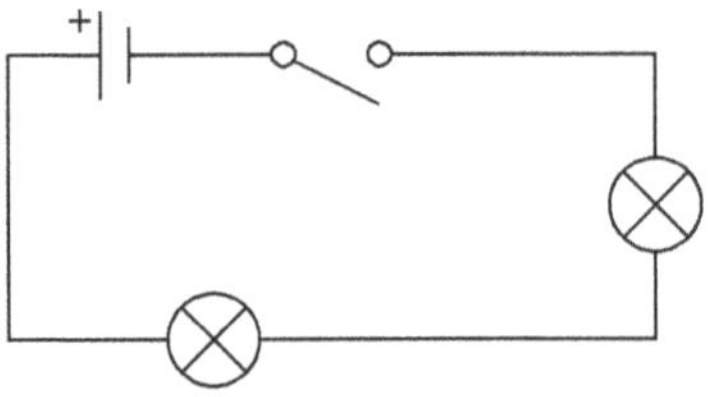

What is the least number of connecting leads the student would need to build the circuit? Tick **one** box.

Three ☐ Four ☐ Five ☐ Six ☐ [1 mark]

3. Look at these four circuit diagrams labelled W, X, Y and Z.

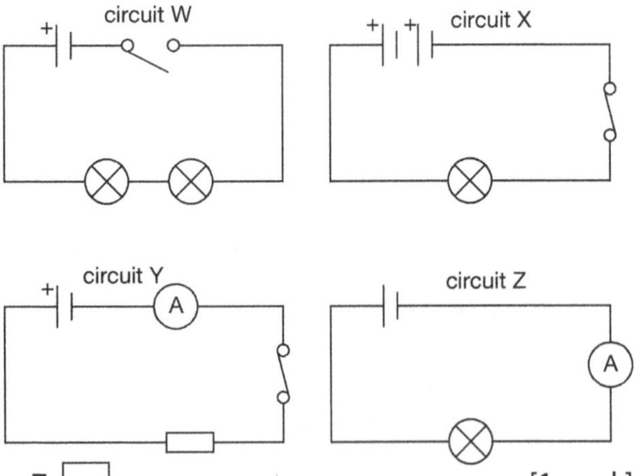

circuit W

circuit X

circuit Y

circuit Z

a Which circuit contains a battery with two cells? Tick **one** box.

W ☐ X ☐

Y ☐ Z ☐ [1 mark]

b Which circuit does **not** contain a switch? Tick **one** box.

W ☐ X ☐ Y ☐ Z ☐ [1 mark]

c Which circuit contains an ammeter positioned to measure the electric current flowing through a resistor? Tick **one** box.

W ☐ X ☐ Y ☐ Z ☐ [1 mark]

d Which circuit contains the symbol for an open switch? Tick **one** box.

W ☐ X ☐ Y ☐ Z ☐ [1 mark]

Electrical charge and current

1. This circuit diagram contains three ammeters, A1, A2 and A3.

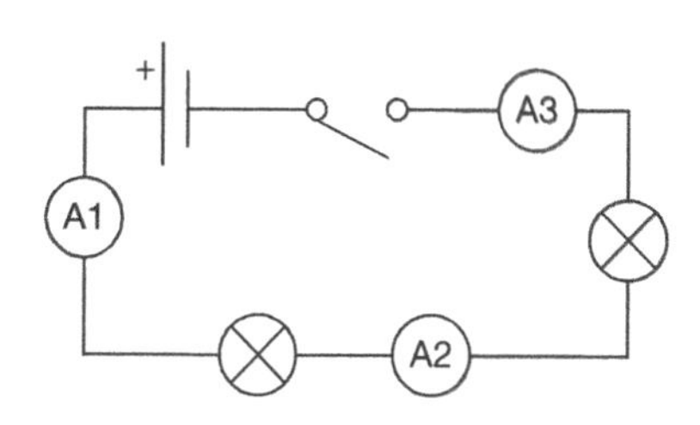

When the switch is closed, are the readings on ammeters A1, A2 and A3 the same or different?

Give a reason for your answer.

____

____ [2 marks]

2. A student builds the circuit shown in the diagram.

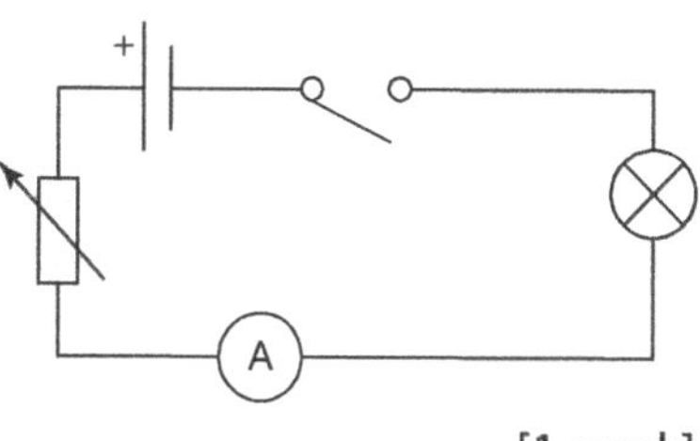

a Complete the sentence.

When the switch is closed, the ammeter measures the rate of flow of ____________________. [1 mark]

b What is the source of potential difference in the circuit?

____ [1 mark]

c Add a suitable meter to the circuit diagram that would enable the potential difference across the lamp to be measured. [2 marks]

Maths

You need to learn the equation for charge flow – it may **not** be given to you in a question.

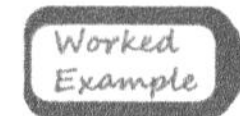

Calculate the current in a wire if a charge of 1.4 C flows through the wire in 4.0 s.

Use the equation

$$\text{charge flow} = \text{current} \times \text{time, or } Q = It$$

Rearrange the equation to make the current, I, the subject. Divide both sides of the equation by time, t.

$$I = \frac{Q}{t}$$

$$= \frac{1.4}{4.0} = 0.35 \text{ A}$$

Maths **d** When the switch is closed, the ammeter reads 0.15 A.

Calculate the charge that flows through the lamp in 20 seconds.

Give the correct unit with your answer.

Charge flow = ______ Unit: ______ [3 marks]

Maths

Electric current values are sometimes given in milliamps (mA). 'milli' is a prefix which means $\frac{1}{1000}$.

So 1 mA is equal to 0.001 A, which can also be written as 1×10^{-3} A.

Maths **e** The variable resistor is now adjusted so that the lamp is much dimmer. The ammeter now reads 20 mA.

How much time does it take for 1.0 C of charge to flow through the lamp?

Time taken = ______ s [2 marks]

Electrical resistance

1. Draw **one** line from each quantity to its correct unit.

Quantity	**Unit**
Potential difference	ampere
Charge	volt
Current	ohm
Resistance	coulomb

[4 marks]

2. Practical

A student wants to measure the resistance of combinations of resistors. He has four identical resistors. Two resistors are connected in parallel and the other two in series.

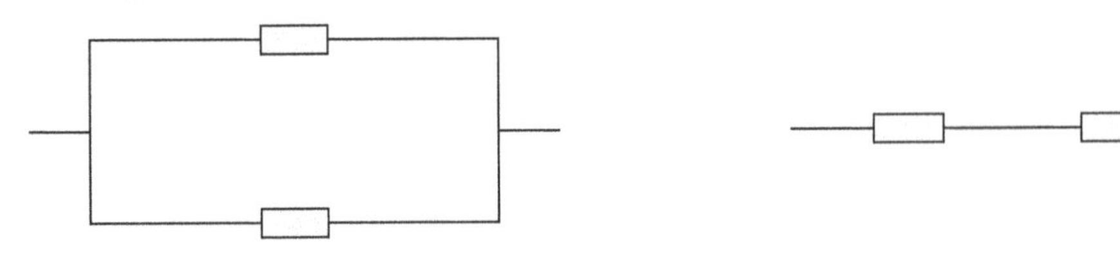

Add components to the correct circuit to show how measurements can be made to determine the combined resistance of two resistors in series. [3 marks]

3. When the switch in the circuit in this diagram is closed, the ammeter reads 0.25 A.

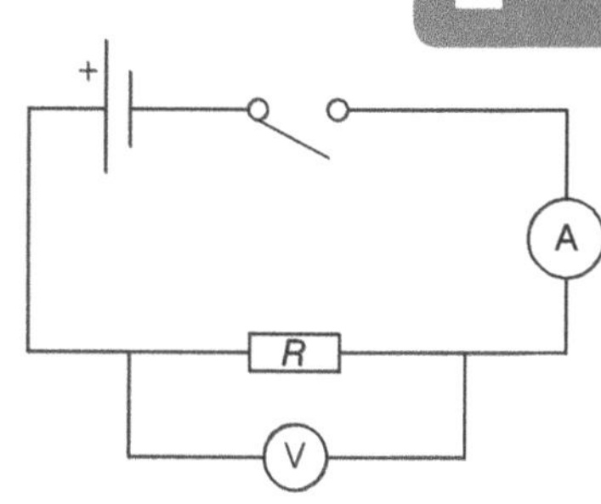

Maths **a** Resistor R has a resistance of 6.0 Ω.

The equation which links current, potential difference and resistance is:

potential difference = current × resistance

Voltmeter reading = ______ V [2 marks]

Maths

You need to learn this equation. It may **not** be given to you in a question.

Calculate the reading on the voltmeter when the switch is closed.

b If resistor R was replaced with a resistor with a much bigger resistance, how would the ammeter reading change?

______________________________ [1 mark]

Maths **c** Resistor R is now replaced with a 1 metre length of nichrome wire. The voltmeter reads 1.5 V and the ammeter reads 48 mA.

What is the resistance of the wire?

Give your answer to 2 significant figures.

Remember

To round a number to two significant figures, look at the third digit. Round up if the digit is 5 or more, and round down if the digit is 4 or less.

Resistance of wire = ______ Ω [3 marks]

4. Practical

A student plans an experiment to find out how the resistance of a piece of constantan wire depends on its length. She builds the circuit shown in the circuit diagram.

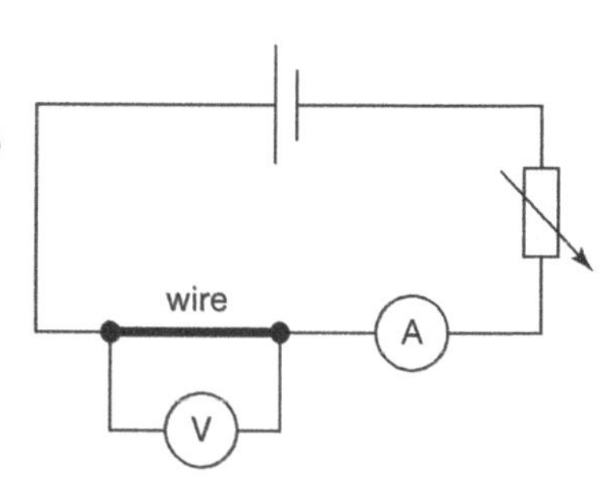

This is the student's plan:

- Attach crocodile clips to the wire.
- Measure the length of wire between the crocodile clips using a metre rule.
- Close the switch.
- Record the readings on the ammeter and voltmeter.
- Repeat the measurements for different lengths of wire.

Maths **a** The student's measurements are shown recorded in the table.

The resistance for the first set of measurements has already been calculated.

Complete the last column. Record the resistance values to the same number of decimal places as the first resistance value. [3 marks]

Length of wire (m)	Current (A)	Potential difference (V)	Resistance of wire (Ω)
0.300	0.443	1.49	3.36
0.400	0.333	1.49	
0.500	0.266	1.49	
0.600	0.222	1.49	
0.700	0.190	1.49	
0.800	0.166	1.49	
0.900	0.150	1.49	

Remember

To get the resistance in ohms, you need to convert the current values to amps.

b The student controlled the type of wire and potential difference across the ends of the wire.

Give **one other** variable that the student should have controlled in her investigation.

______________________________ [1 mark]

c Give **two** reasons why taking repeat readings for each length of wire would improve the experiment.

____

______________________________ [2 marks]

Maths **d** The student plots a line graph to analyse the measurements.

Which variable should go on each axis?

x-axis: ______________________________

y-axis: ______________________________ [2 marks]

Practical

When plotting a graph to analyse experiment measurements, the independent variable is usually plotted on the *x*-axis.

e To prevent the wire from getting too hot, the student used only one cell in the circuit.

Suggest another way of making sure the wire does not get too hot.

______________________________ [1 mark]

Resistors and *I–V* characteristics

1. Practical

A student is given a resistor labelled 6.8 Ω. She is asked to build a circuit to test whether the resistance of the resistor remains 6.8 Ω as the current through it is changed.

She is given this incomplete circuit diagram. The diagram has a component missing between points X and Y.

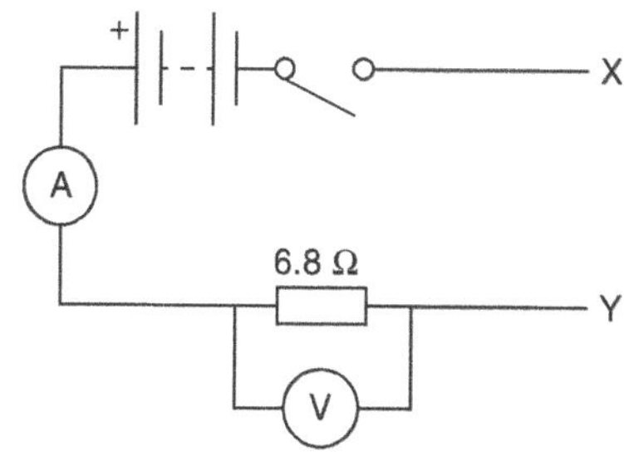

a Name a component that the student can adjust to vary the size of the current in the circuit.

______________________ [1 mark]

b Complete the circuit diagram above to show the correct circuit diagram symbol for this component, between points X and Y. [1 mark]

c The student made a series of current and potential difference measurements for the 6.8 Ω resistor. She then plotted this graph of current against potential difference.

What change would she have made to the circuit to obtain the negative current and potential difference values?

____

______________________ [1 mark]

d What does the graph show about the relationship between the current and the potential difference?

Explain your answer.

____

____

______________________ [3 marks]

2. Practical

a Complete the sentence. Use words from the box.

potential difference	resistance	charge	temperature

A component is described as an ohmic conductor if its ______________________

stays constant as the current through it is changed, at constant

______________________. [2 marks]

b This is a graph of current against voltage for a diode. The graph shows that current flows through the diode in one direction only.

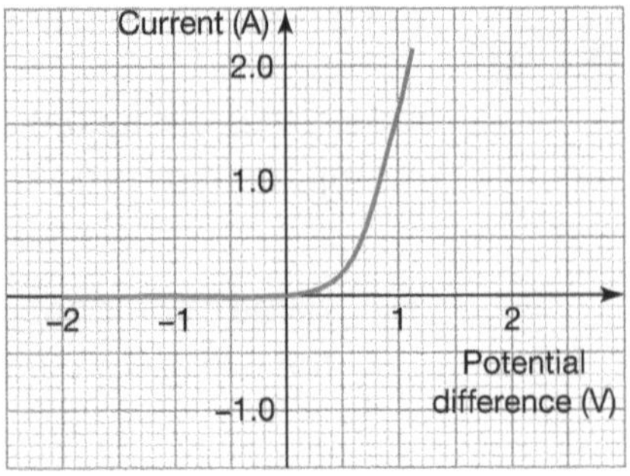

What is the relationship between the current and the potential difference in this graph? Tick **one** box.

Directly proportional ☐ Non-linear ☐

Inversely proportional ☐ [1 mark]

Current (A)
2.0
1.0
−2
−1
1
2
−1.0
Potential difference (V)

c Describe the resistance of the diode in the reverse direction.

______________________ [1 mark]

3. Practical

A box labelled X contains a mystery electrical component. A student is asked to investigate the relationship between the current through the component and the potential difference applied across it.

Design a method the student could use to investigate the relationship between the current and potential difference for the mystery component.

Include in your description the measurements she should make and how she should obtain negative values.

A circuit diagram may be drawn as part of your answer.

> **Practical**
> When the question asks you to **design** an experiment, you need to set out the steps – how the experiment will be done. Picture yourself doing the experiment so that the instructions will be clear to someone else even if they have not done a similar experiment before.

____

____

____

______________________ [6 marks]

4. This circuit contains a light-dependent resistor (LDR).

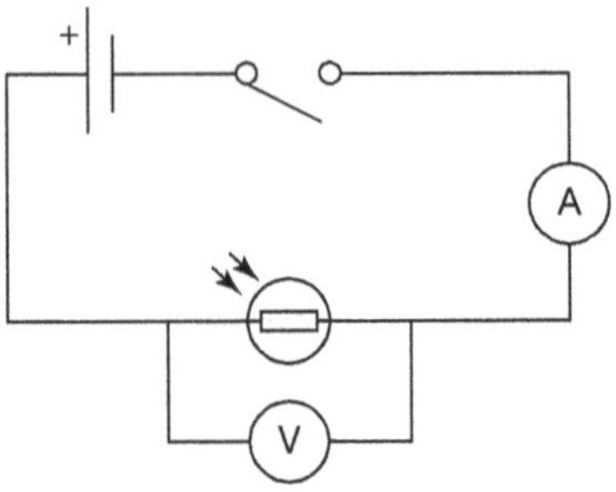

+
A
V

a What quantity does the voltmeter measure?

______________________ [1 mark]

Maths **b** When the switch is closed the meter readings are taken. The ammeter reads 3.52 mA and the voltmeter reads 1.51 V.

Calculate the resistance of the LDR.

Resistance = ______ Ω [2 marks]

Maths **c** The LDR is now covered with a black cloth that stops any light from hitting the LDR. The ammeter now reads 1.23 mA and the voltmeter reads 1.51 V.

Calculate the new resistance of the LDR.

Resistance = ______ Ω [2 marks]

d What can you conclude about how the resistance of an LDR can be changed?

______

______ [2 marks]

5. This diagram shows an ohmmeter used to measure the resistance of an electrical component.

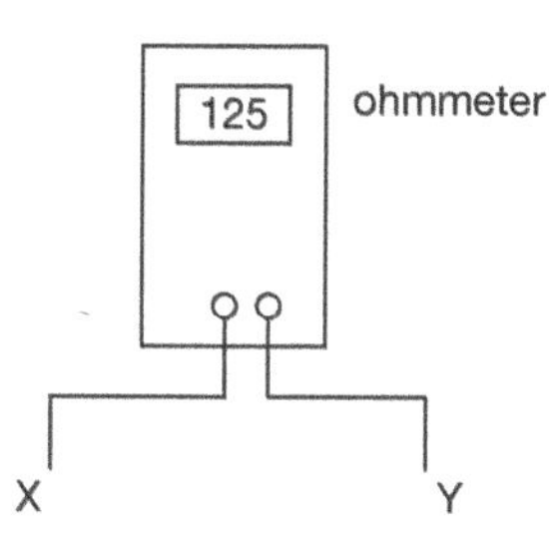

a A student is asked to measure the resistance of a thermistor using an ohmmeter.

Complete the circuit diagram to show the thermistor connected between points X and Y. [1 mark]

b The thermistor is placed in a beaker of hot water to change its temperature. The resistance of the thermistor is measured as the water cools down. The water temperature is measured with a thermometer.

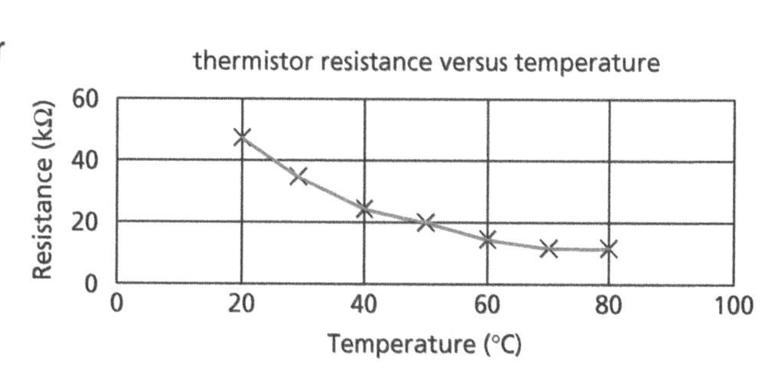

The student plots a graph of resistance against temperature.

What can you conclude from this graph about how the resistance of a thermistor can be changed?

______

______ [2 marks]

Series and parallel circuits

1. **a** This circuit diagram contains three identical resistors.

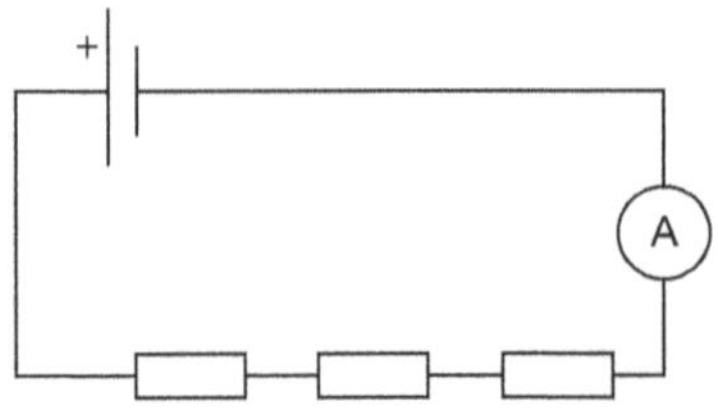

Which statements about the circuit are true?
Tick **two** boxes.

The resistors are in series. ☐

Each resistor has a different current through it. ☐

The same current flows through all three resistors. ☐

The resistors are in parallel. ☐ [2 marks]

Maths **b** Each of the resistors has a resistance of 2.2 Ω.

What is the equivalent resistance of the three resistors in this circuit?

Equivalent resistance = ______ Ω [1 mark]

Maths **c** The potential difference across the cell is 1.5 V.

What will the potential difference across one of the 2.2 Ω resistors be when the switch is closed?

Potential difference = ______ V [1 mark]

2. Three different resistors are connected to a cell as shown in the circuit diagram.

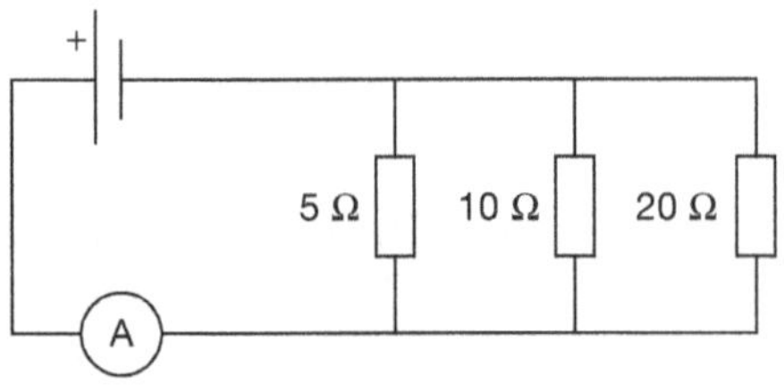

5 Ω 10 Ω 20 Ω

a Which statements apply to the circuit above?
Tick **two** boxes.

The resistors are in series. ☐

Each resistor has a different potential difference across it. ☐

The three resistors have the same potential difference across them. ☐

The resistors are in parallel. ☐ [2 marks]

Maths **b** The potential difference across the cell is 1.5 V.

Calculate the current through the 5 Ω resistor.

Current = ______ A [2 marks]

3. Maths

In this circuit diagram, the potential difference across the cell is 1.5 V.

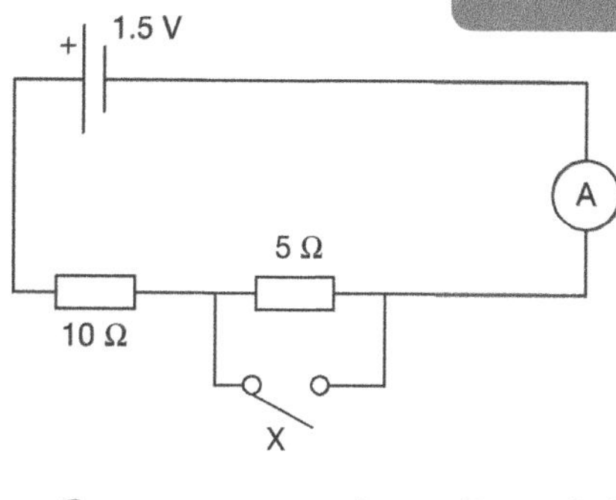

a Calculate the current flowing through the ammeter when switch X is open.

Current = _____ A [2 marks]

b Calculate the current flowing through the ammeter when switch X is closed.

Current = _____ A [2 marks]

Mains electricity

1. **a** What is the frequency of the mains electricity supply in the UK?

_____ [1 mark]

b What is the potential difference of the mains electricity supply in the UK?

_____ [1 mark]

2. Draw **one** line from each wire in a mains electricity cable to the job that it does.

Wire	**Job done**
Neutral	Carries the alternating potential difference from the supply
Live	Acts as a safety device
Earth	Completes the circuit

[3 marks]

Common misconception
Electricity is not an energy store, but a way of transferring energy from one store to another.

3. Complete the sentences to describe the mains electricity supply in the UK. Use words from the box.

direct	live	earth	alternating	blue	brown	neutral

The potential difference of the UK mains electricity supply is ______________________.

The live wire in the mains cable is coloured ______________________.

If the appliance becomes faulty, the ______________________ wire prevents the person using the appliance from receiving an electric shock. [3 marks]

4. An electric kettle can develop a fault in which the outer metal casing becomes connected to the live wire inside the kettle.

Explain how the earth wire prevents anyone who touches the kettle from getting an electric shock.

____

____

____ [3 marks]

Command words

There are three marks here so make three points. Remember that '**explain**' means you have to say **what** happens and also **why** it happens.

Energy changes in circuits

1. Complete the sentences to describe the energy transferred by the appliances. Choose your answers from the box.

thermal	gravitational potential	kinetic

A food mixer transfers energy from the mains electricity supply to ______________________ energy. The current in an electric kettle transfers energy to water as ______________________ energy. [2 marks]

2. A vacuum cleaner transfers energy from the mains electricity supply to air, increasing its store of kinetic energy.

Which changes would result in a bigger transfer of energy by a vacuum cleaner?
Tick **two** boxes.

Operate the vacuum cleaner for a longer period of time. ☐

Operate the vacuum cleaner for a shorter period of time. ☐

Use a vacuum cleaner with a lower power rating. ☐

Use a vacuum cleaner with a higher power rating. ☐ [2 marks]

3. Maths

An electric heater which can be plugged into the mains electricity supply has a power rating of 2000 W.

Calculate the energy store the heater transfers in 30 minutes.

Use the equation

$$\text{energy transferred} = \text{power} \times \text{time}$$

Maths
You need to learn this equation – it may **not** be given to you in a question.

Energy store transferred = _______________ J [2 marks]

Worked Example

An electric hair dryer operates at a power of 1200 W when plugged into the 230 V mains electricity supply.

Calculate the total charge that has passed through the hair dryer when it has transferred 60 000 J of energy.

Use the equation

$$E = QV$$

Rearrange to give $Q = \frac{E}{V}$

$Q = \frac{60\,000}{230} = 261\ \text{C}$

Maths
You need to learn this equation – it may **not** be given to you in a question.

4. Maths

A microwave oven used in a café has a power of 1.0 kW. It takes 5 minutes to cook a jacket potato when connected to the 230 V mains electricity supply.

a Calculate the energy store transferred by the microwave oven in 5 minutes.

Energy store transferred = _______________ J [2 marks]

b Calculate the total charge that flows from the mains electricity supply to the microwave oven during the 5 minutes it is in use.

Charge = _______________ C [2 marks]

5. Maths

An LED torch operates at a power of 2.0 W when connected to a 3.0 V battery.

a Calculate the energy store transferred by the torch in 10 minutes of use.

Energy store transferred = _______________ J [2 marks]

b Calculate the total charge that flows through the torch in 10 minutes of use.

Charge = ______ C [2 marks]

Electrical power

1. Maths

Calculate the power of an electric food mixer that carries a current of 2.0 A when plugged into the 230 V mains electricity supply.

Use the equation

power = potential difference × current

Give the correct unit for your answer.

Maths

You need to learn this equation – it may **not** be given to you in a question.

Power = ______ Unit: ______ [2 marks]

Worked Example

The current in a resistor transfers energy to the resistor as thermal energy. The resistor gets hot.

Calculate the energy transferred by second by a 6.0 Ω resistor if it was carrying a current of 80.0 mA.

Power, $P = I^2R$

$= 0.08^2 \times 6.0 = 0.49$ W

2. Maths

A 30.0 m long extension cable has a resistance of 0.40 Ω.

Calculate the power loss in the extension cable when it is carrying a current of 4.0 A.

Power loss = ______ W [3 marks]

3. Maths

Microwave oven A draws a current of 2.6 A when connected to the 230 V mains electricity supply. Microwave B draws a current of 3.9 A when connected to the 230 V mains electricity supply.

Explain why microwave B cooks food more quickly than microwave A.

____

____

____ [3 marks]

4. An electrical heater for a fish tank draws a current of 0.40 A when connected to the 230 V mains electricity supply.

Maths

a Calculate the power of the heater.

Power = ______ W [2 marks]

b Calculate the energy store transferred by the heater when it is switched on for 10 minutes.

Energy store transferred = __________ J [2 marks]

The National Grid

1. The National Grid system delivers electricity generated at power stations to towns and cities all over the country.

Give **one** reason why the National Grid system is better than each town having its own power station.

____

____ [1 mark]

2. Complete the sentences about transformers. Choose your answers from the box.

increases	decreases

A step-up transformer ____________________ the potential difference and ______________ the current. A step-down transformer ____________________ the potential difference and ____________________ the current. [4 marks]

3. In the diagram, the cables shown as lines are part of the National Grid system. Transformer A is connected between the power station and the National Grid. Transformer B is connected between the National Grid and a school.

Which is the correct description of transformers A and B? Tick **one** box.

Transformer A is a step-down and transformer B is a step-up. ☐

Transformer A and transformer B are both step-up. ☐

Transformer A is a step-up and transformer B is a step-down. ☐

[1 mark]

Synoptic

More information about how transformers work is included in the Higher Tier work in section 7.

4. Which two statements about the National Grid are correct? Tick **two** boxes.

The National Grid transmits electricity at high voltages to improve efficiency. ☐

The National Grid transmits electricity at low voltages to improve efficiency. ☐

The National Grid transmits electricity at high currents to improve efficiency. ☐

The National Grid transmits electricity at low currents to improve efficiency. ☐

[2 marks]

Static electricity

1. Two light spheres are found to repel each other when suspended.

Which two statements could be correct explanations for the repulsion between
the two spheres. Tick **two** boxes.

Sphere A is positive, sphere B is positive. ☐

Sphere A is positive, sphere B is uncharged. ☐

Sphere A is negative, sphere B is negative. ☐

Sphere A is uncharged, sphere B is negative. ☐

[2 marks]

2. **a** A polythene rod becomes negatively charged when it is rubbed with a cloth. Describe how this happens.

______________________________ [1 mark]

b Explain what effect rubbing the rod has on the cloth.

____

______________________________ [2 marks]

Command words

In an **explain** question, if there are 2 marks, make 2 clear points. The second mark is for giving a **reason** for what happens.

c When an acetate rod is rubbed with a cloth it becomes positively charged.

Describe what happens when a charged acetate rod is moved towards a suspended charged polythene rod.

______________________________ [1 mark]

d Explain your answer to part **c**.

____

______________________________ [2 marks]

Electric fields

1. **a** A **negatively** charged metal sphere is shown in the diagram.

The electric field around the charge is represented by lines but the diagram is incomplete. Add arrows to show the direction of the electric field. [1 mark]

b A small negatively charged object is placed in the electric field created by the charged metal sphere.

What happens to the size of the force on this charged object as it gets closer to the metal sphere? Tick **one** box.

Force gets bigger ☐ Force is constant ☐ Force gets smaller ☐ [1 mark]

2. What would be the direction of the force exerted by the electric field in the diagram above in question **1** on a positively charged speck of dust in the air? Tick **one** box.

Towards the sphere ☐   Away from the sphere ☐   [1 mark]

3. The underneath of a thunder cloud becomes negatively charged. The potential difference between the bottom of the cloud and the ground is about 5 million volts. The lightning strike carries a charge of 15 C.

cloud
potential difference
5 MV
lightning strike
ground

a About how much energy store is transferred as the lightning strikes? Tick **one** box.

7.5 MJ ☐   75 MJ ☐   750 MJ ☐   7500 MJ ☐   [1 mark]

b The lightning strike lasts 500 μs. What current flows during the lightning strike?
Tick **one** box.

30 A ☐   300 A ☐

3000 A ☐   30 000 A ☐   [1 mark]

Maths
Short time intervals are sometimes given in microseconds (μs). 'micro' is a prefix equal to $\frac{1}{1\,000\,000}$.
So 1 μs can be written as 1×10^{-6} s.]

c What is the resistance of the air along the track taken by the lightning? Tick **one** box.

17 Ω ☐   170 Ω ☐   1700 Ω ☐   17 000 Ω ☐   [1 mark]

Density

1. Write down the equation which links density, mass and volume.

______________________________ [1 mark]

Maths

You need to learn this equation – it may not be given to you in a question.

Worked Example

Calculate the mass of an object with a density of 2 kg/m^3 and volume of 3 m^3.

Use the equation density = $\frac{\text{mass}}{\text{volume}}$

Rearrange the equation to make mass the subject.

mass = density × volume

= 2 kg/m^3 × 3 m^3

= 6 kg

2. Maths

A wooden object has a density of 740 kg/m^3. Its mass is 3700 g.

Calculate the object's volume.

Give the unit.

Volume = ______ Unit ______ [4 marks]

Maths

In this example you'll need to rearrange the equation to find the mass. This means putting mass on its own, on one side of the equals sign.

This equation is given in the form $a = b/c$. To find b, multiply both sides by c.

Maths

Be careful of units when you substitute values into an equation. The unit of mass needs to match the unit of density. There are 1000 g in 1 kg.

3. Practical

a Which apparatus would be best for measuring a length less than 5 mm in size? Tick **one** box.

Metre ruler ☐ Micrometer ☐ 30 cm ruler ☐ [1 mark]

b Which apparatus is used for measuring the volume of a liquid? Tick **one** box.

Measuring cylinder ☐ Displacement can ☐ 30 cm ruler ☐ [1 mark]

4. Practical

a A scientist finds a rock they believe to be a small meteorite. The rock has an **irregular** shape.

Describe a method the scientist could use to calculate the density of the rock.

Your answer should consider possible sources of error and how these could avoided.

Practical

When the question asks you to **describe a method**, you need to make the steps clear. You could use bullet points or subheadings. Make sure you link the points you make about **sources of error** to the apparatus that is used and how it should be used.

[6 marks]

Maths

b Most meteorites have a density greater than most terrestrial (Earth) rocks.

Density of meteorites: 3000 to 8000 kg/m^3

Density of terrestrial rocks 2500 to 3200 kg/m^3

The mass of the rock is found to be 0.112 kg and the volume is 0.000 031 m^3.

Show that the rock is probably a meteorite.

[4 marks]

5. Describe a method you could use to find the density of an unknown liquid in the laboratory.

Higher Tier only

Synoptic

____

____

____

____________________ [4 marks]

Synoptic

In the final exam some of the marks will be for connecting your knowledge from different areas of physics. To answer this question you need to link knowledge of the particle model with ideas about upthrust (section 5).

The diagram shows the arrangement of particles in a liquid. Each circle represents one particle.

a Draw a diagram in the box to show the arrangement of particles in a gas.

[2 marks]

b Use the diagrams to explain why gas bubbles accelerate upwards in fizzy drinks.

____

____

____________________ [3 marks]

Changes of state

1. a Draw **one** line to match each change of state with its description. [4 marks]

Change of state	Description
Freeze	Gas changes to liquid
Boil	Solid changes to liquid
Condense	Liquid changes to gas
Melt	liquid changes to solid

b For which **two** changes of state does energy needs to be **supplied**?

____

____ [2 marks]

2. Iodine is a material which sublimates.

What is meant by 'sublimates'?

____ [1 mark]

3. What makes freezing a **physical** (**not** chemical) change?

____ [1 mark]

4. Which one of the following is **not** a physical change? Tick **one** box.

Sublimation ☐ Burning ☐ Evaporating ☐ [1 mark]

5. A liquid is poured into a tray and then frozen.

Student A thinks that the mass of the frozen solid will be greater than the mass of the liquid.

Student B thinks that the masses will be the same.

Who do you agree with?

Explain your answer.

____

____ [2 marks]

Specific heat capacity and latent heat

1. Describe the movement of the particles in a hot gas compared to a cold gas.

____ [1 mark]

2. When a gas is heated, which energy store in the gas increases? Tick **one** box.

Kinetic energy ☐ Gravitational potential energy ☐ Elastic energy ☐ [1 mark]

3. There is a bath of luke-warm water next to a cup of hot coffee.

a The coffee has a higher temperature. What does this tell you about the particles it contains?

______________________________ [1 mark]

b The water in the bath has a greater store of internal energy than the coffee in the cup.

Explain why, in terms of the particles it contains.

____

______________________________ [2 marks]

4. 1 kg of ice at 0 °C has less internal energy than 1 kg of water at 0 °C. Explain why.

____

______________________________ [2 marks]

5. A block of ice is heated for a few minutes. Even though thermal energy is still being supplied, the temperature of the ice and the melted water does not increase. Explain why.

____

____

______________________________ [2 marks]

6. Look at the graph.

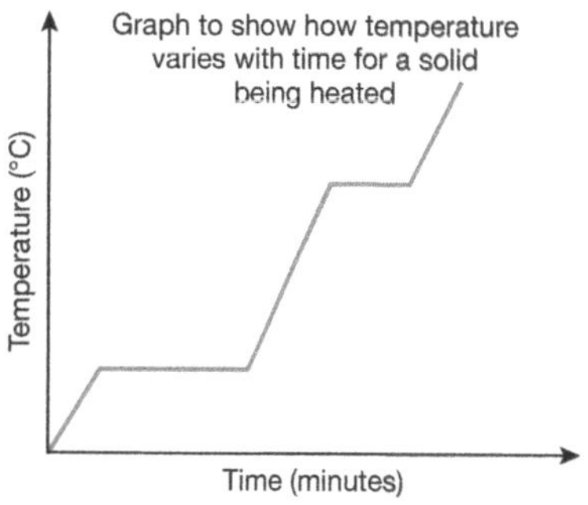

a Label the following regions on the graph: **melting**, **liquid**, **boiling**, **solid**. [4 marks]

b Which of these regions does the term **specific latent heat of fusion** relate to?

______________________________ [1 mark]

Maths **c** The specific latent heat of fusion for this material is 45 kJ/kg.

Calculate the energy required to change 5 kg of the material from a solid to a liquid.

Use the equation

energy for a change of state = mass × specific latent heat

Maths

This equation is provided in the Physics Equation Sheet. You should be able to select and apply the equation – it may **not** be given in the question.

Energy = ______ J [2 marks]

7. A 500 g sample of a material is found to require 15 000 J to melt.

Calculate the specific latent heat of fusion for this material.

Use the correct equation from the Physics Equation Sheet.

Maths

You will need to select the correct equation, then rearrange the equation to give specific latent heat as the subject. Try dividing both sides of the equation by mass.

Specific latent heat of fusion = ______ J/kg [3 marks]

Particle motion in gases

1. Why does a gas exert pressure on its container?

______________________________ [1 mark]

2. What changes about the particles in a gas when the temperature is increased?

______________________________ [1 mark]

3. The temperature of a gas in a container of fixed volume is increased.

Explain what will happen to the pressure.

Command word

'Explain' means you have to say **what** happens and also **why** it happens. There are 3 marks here so make **three** points.

____

____

______________________________ [3 marks]

4. It is dangerous to heat a sealed conical flask over a Bunsen burner.

Synoptic

a Describe **one** possible dangerous outcome.

____ [1 mark]

b Explain your answer.

____

____ [2 marks]

5. A cook pours hot jam into a jar. He immediately puts the lid on, leaving a layer of air above the jam's surface. After the jam has cooled, it becomes difficult to remove the lid.

Synoptic

Explain why.

Literacy

Remember to use **key words** in your explanation. **Temperature** and **pressure** are two terms we have used a lot in this topic so try to include them. You will need to use any information given to you, and not just repeat it. Think about how to link words and phrases.

____

____

____ [4 marks]

Increasing the pressure of a gas

1. A sealed container has a toxic gas inside at high pressure. The owners are concerned it will burst. However, they are able to increase the volume of the container without letting any gas in or out.

Synoptic

Explain why this will help.

A diagram may be drawn as part of your answer.

____

____ [4 marks]

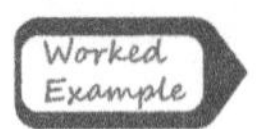

A helium balloon expands in volume as it rises in the atmosphere. It starts with a volume of 0.02 m^3 at a pressure of 100 000 Pa.

Calculate the new volume when the pressure reduces to 50 000 Pa.

Use the equation

pV = constant

pV before = 0.02 × 100 000 = 2000

pV after = 2000 too, therefore

V after = 2000/50 000

= 0.04 m^3

Maths

This equation is given in the Physics Equation Sheet. But remembering the version: $p_1V_1 = p_2V_2$ may be more helpful. The subscript '1' refers to 'initial' and '2' to 'final'.

2. *Maths*

A bike tyre is initially empty of any air. It is then pumped up by compressing 0.001 m^3 of air at atmospheric pressure. The final volume of the tyre is 0.0002 m^3.

Calculate the final pressure in the tyre.

Use the equation: pV = constant

Maths

This question has more than one step. Always show your working out. You may still get some marks, even if your final answer is wrong.

Pressure = ______ Pa [5 marks]

3. *Higher Tier only* *Synoptic*

The equation you used in question **2** is only valid if the **temperature** of the gas has been kept constant.

Is this likely? Explain your answer.

____

____

______________________________ [2 marks]

Protons, neutrons and electrons

1. Complete the table to describe the properties of protons, neutrons and electrons.

The electrons in an atom occupy different energy levels.

Particle	Charge	Mass	Located inside or outside nucleus?
	positive		
neutron			
		negligible	outside nucleus

[8 marks]

2. Complete the sentences.

The energy levels correspond to how far the electrons are from the ____________________.

Higher energy levels are ____________________ from the nucleus. Electrons can move between energy levels – if an electron emits electromagnetic radiation it has moved to a ____________________ energy level. [3 marks]

3. The Sun emits radiation at all wavelengths in the electromagnetic spectrum. To reach the surface of the Earth, this radiation must pass through the gases in the Sun's outer layers and the gases in the Earth's atmosphere.

Suggest what happens to atoms in these gases when radiation from the Sun passes through.

____

____ [2 marks]

The size of atoms

1. Maths

Put these powers of 10 in order of size from **smallest** to **largest**:

10^3	10^{-5}	10^1	10^4

Maths

Remember, a **negative** power of ten means a number **less than one.**

____________________________ [4 marks]

2. Maths

Which one of these measurements is equivalent to 0.002 m?
Tick **one** box.

2×10^3 m ☐ 2×10^{-2} m ☐

2×10^2 m ☐ 2×10^{-3} m ☐ [1 mark]

3. If the nucleus of an atom was the size of an apple, approximately what would the size of the whole atom be? Tick **one** box.

A plate ☐ An armchair ☐ A room ☐ A football pitch ☐ [1 mark]

4. Look at the distances below.

A 1×10^{-10} m

B 1×10^{-6} m

C 1×10^{-14} m

D 1×10^{-20} m

Which is closest to:

a The radius of an **atom**?

____________________ [1 mark]

b The radius of a **nucleus**?

____________________ [1 mark]

Elements and isotopes

1. Synoptic

Complete the sentences.

Use words from the box.

positive	negative	neutral	electrons
neutrons	protons	mass	atomic

Atoms have a ____________________ charge. Therefore the number of protons must be equal to the number of ____________________. An element is defined by the number of ____________________ in the nucleus. This is known as the element's ____________________ number. The mass number is the sum of the number of ____________________ and number of ____________________. [6 marks]

2. Complete the table.

Atomic number of element	Mass number	Number of protons	Number of neutrons
	12	6	
11			12
	59		32
		13	14

Literacy
Two of these column headings basically mean the **same** thing. Which are they?

[8 marks]

3. Define an isotope.

____

____

______________________________ [3 marks]

4. An isotope of cobalt has a mass number of 60. The atomic number of cobalt is 27.

a Write down the number of protons, neutrons and electrons in an atom of this isotope.

Number of protons = ______

Number of neutrons = ______

Number of electrons = ______ [3 marks]

b The chemical symbol for cobalt is Co.

Write in symbol form how the nucleus of this isotope can be represented.

______ [1 mark]

c Another isotope of cobalt has a mass number of 59.

Write in symbol form how the nucleus of this isotope can be represented.

______ [1 mark]

Electrons and ions

1. Draw **one** line to match each statement on the left to the correct result on the right.

Statement	**Results in**
Having a different number of neutrons in the nucleus	An ion of the same element
Having a different number of electrons orbiting the nucleus	A different element
Having a different number of protons in the nucleus	A different isotope of the same element

[3 marks]

2. An atom **loses** an electron.

Is it now positively charged or negatively charged?

Explain why.

____

____ [3 marks]

3. An atom of chlorine has a mass number of 35 and the atomic number is 17. The atom gains an electron to become an ion with a charge of –1.

Give the number of protons, neutrons and electrons in the ion.

Number of protons = ______

Number of neutrons = ______

Number of electrons = ______ [3 marks]

4. An atom of aluminium has a mass number of 27 and the atomic number is 13. The atom loses two electrons to become an ion with a charge of +2.

Give the number of protons, neutrons and electrons in the ion.

Number of protons = ______

Number of neutrons = ______

Number of electrons = ______ [3 marks]

Discovering the structure of the atom

1. The earliest idea about atoms was that atoms are the smallest particle.

State the discovery which showed this theory to be incorrect.

_______________ [1 mark]

2. **a** Describe the plum pudding model of the atom.

____

_______________ [2 marks]

b Which statement best describes where positive charge is located in the plum pudding model?

Concentrated in a small space ☐ Spread out everywhere ☐ [1 mark]

3. Describe the differences between Rutherford's model of the atom and Niels Bohr's model of the atom.

____

_______________ [2 marks]

4. The alpha-scattering experiment showed that the nucleus was charged.

State the **two** later discoveries that gave further information about the structure of the nucleus.

____

_______________ [2 marks]

5. Look at the diagram of the alpha-scattering experiment.

a Label the grey box and gold square. [2 marks]

b What results would be expected in this experiment if the plum pudding model is correct?

Tick **one** box.

All the alpha particles would bounce back. ☐

All the alpha particles would go straight through. ☐

Some alpha particles would go through and others would bounce back. ☐

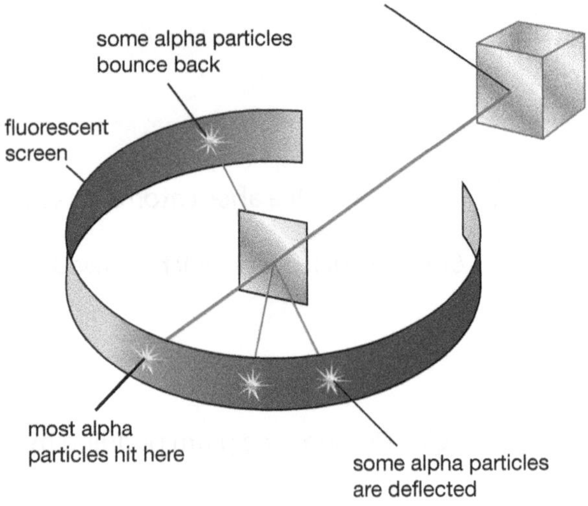

[1 mark]

c Describe the results actually obtained in the alpha-scattering experiment.

____

____ [2 marks]

6. Ernest Rutherford developed a new model of the atom from the results of the alpha-scattering experiment.

Describe the Rutherford model of the atom.

Explain how the results of the alpha-scattering experiment provided evidence for this model.

Literacy

To break this question down, take each result in turn and explain what it implies about the structure of the atom.

____

____

____

____

____

____

____

____ [6 marks]

Radioactive decay

1. Name the subatomic particle that is lost from an atom by beta decay.

_______________ [1 mark]

2. Complete the sentences.

Nuclear radiation refers to radiation emitted from the _______________ of an atom. This process is known as radioactive decay and occurs when the nucleus is _______________. It is impossible to predict exactly when a nucleus will decay – it is a _______________ process. The activity of a radioactive source is a measure of its rate at which the nuclei decay and is measured in _______________. Geiger–Muller tubes detect the number of decays per second, which is known as the _______________. [5 marks]

3. When an atom decays it emits one of four types of nuclear radiation: an alpha particle, a beta particle, a gamma ray or a neutron.

a In which type of radioactive decay is a neutral particle emitted?

_______________ [1 mark]

b Which type of nuclear radiation consists of a high-speed electron?

_______________ [1 mark]

c Which type of nuclear radiation is an electromagnetic wave?

_______________ [1 mark]

d Describe the structure of an alpha particle.

____

_______________ [2 marks]

e Describe what happens when a radioactive nucleus emits a beta particle.

____

_______________ [2 marks]

Comparing alpha, beta and gamma radiation

1. Complete the table to compare the properties of alpha, beta and gamma radiation.

Type of radiation	Stopped by	Range in air
beta		
	several cm of lead	
		1–2 cm

[6 marks]

2. Alpha, beta and gamma radiation are all forms of **ionising** radiation.

a What is meant by **ionising**?

______________________________ [1 mark]

b Which type of nuclear radiation is the most ionising?

______________ [1 mark]

3. To diagnose problems with organs in the body, a radioactive tracer can be injected into a patient. The radiation emitted by the tracer is then detected outside the patient's body.

Give **one** reason why a beta source is not suitable to use as a tracer in the body.

______________________________ [1 mark]

4. Ionising radiation is used to kill cancerous cells.

a Would you use an alpha or beta source to kill cancerous cells near the surface of the skin? Give a reason for your answer.

____

____

______________________________ [1 mark]

b Explain why gamma radiation is not suitable to use for treating skin cancer.

____

______________________________ [2 marks]

5. One of the uses of nuclear radiation is to monitor the thickness of paper being produced in a factory.

Explain which of alpha, beta and gamma radiation you would choose for this application.

____

____

______________________________ [3 marks]

Radioactive decay equations

1. Which symbol correctly represents an alpha particle? Tick **one** box.

$^{4}_{4}He$ ☐ $^{4}_{0}He$ ☐ $^{4}_{2}He$ ☐ $^{2}_{2}He$ ☐ [1 mark]

Remember

To help you remember the symbol for an alpha particle, remember that the alpha particle is identical to a helium nucleus. The question is asking you to show the mass and atomic number for a helium nucleus.

2. Which symbol correctly represents a beta particle? Tick **one** box.

$^{1}_{0}e$ ☐ $^{0}_{1}e$ ☐ $^{0}_{-1}e$ ☐ $^{-1}_{-1}e$ ☐ [1 mark]

Remember

The fact an electron has a negative atomic number is a bit confusing! Remember that when a beta particle is emitted a neutron turns **into** a proton. The total atomic number must be the same before and after.

Worked Example

An atom of the isotope polonium-84 decays into an atom of lead-82.

State whether this is alpha or beta decay.

Complete the nuclear equation.

$^{208}_{84}Po \rightarrow \_\_\__{82}Pb + \_\_\_$ [2 marks]

The decay shows that two protons have been lost from the nucleus that decays. This must mean it is an alpha decay and so the missing particle on the right is $^{4}_{2}He$.

The mass numbers have to balance either side of the equation. An alpha particle has a mass number of 4 so the mass number of the lead (Pb) isotope must be equal to 208 – 4 = 204.

3. Complete the nuclear equation below to represent the beta decay of carbon-14.

$^{14}_{6}C \rightarrow$ ____ ^{14}N + ____ e [2 marks]

4. An atom of thorium-90 decays into an atom of radon (Ra) by emitting an alpha particle.

Complete the nuclear equation below to represent the decay of thorium-90.

$^{229}_{90}Th \rightarrow$ ____ $_{88}Ra$ + ____ [2 marks]

5. An atom of the isotope lead-82 decays into an atom of bismuth (Bi) by emitting a beta particle.

Complete the nuclear equation for this decay.

$^{209}_{82}Pb \rightarrow$ ____ Bi + beta particle [2 marks]

6. An atom of cobalt-27 decays into an atom of nickel. Nickel has an atomic number of 28.

Remember

Both the mass numbers and the atomic numbers have to balance on both sides of the equation.

a What particle has been emitted by this radioactive decay?

______________________________ [1 mark]

b Complete the nuclear equation for this decay.

$^{60}_{27}Co \rightarrow$ ____ Ni + ____ [2 marks]

Half-lives

1. What is the half-life of a radioactive isotope? Tick **one** box.

The average time it takes for the mass of the sample to halve. ☐

The average time it takes for the activity of the sample to halve. ☐

The average time it takes for the mass number to halve. ☐ [1 mark]

Common misconception

Note that 'mass of the sample' is not the same as 'mass of the radioactive isotope'. When the isotope decays, it still has mass. So the **total** mass of the sample may **not change** very much, but the mass of the original radioactive isotope will decrease a lot.

2. The graph shows the number of counts recorded per second from a sample of radioactive material.

Maths

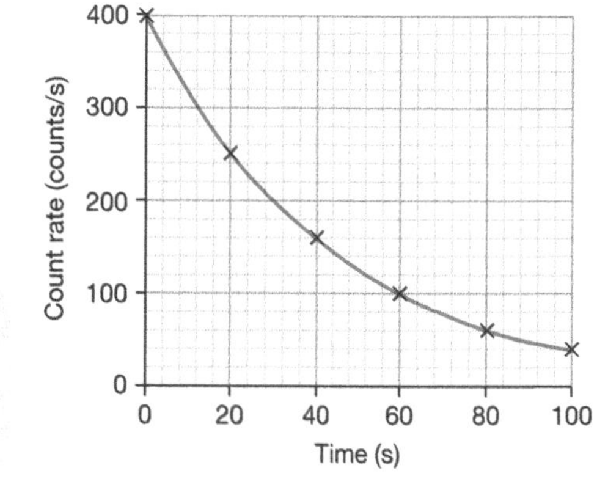

a Use the graph to determine the half-life of this material.

Command words

The question asks you to **show** how you work out your answer from the graph. This means you must draw a line on the graph to show how you worked out your answer.

Show on the graph how you work out your answer.

_____ seconds [2 marks]

b Determine how long it will it take for the count rate to reduce from its starting value of 400 counts/s to approximately 6 counts/s.

Maths

Try writing down 400 then drawing an arrow. Then write down the number which is half of 400. Then draw another arrow. Repeat this until you reach about 6. Then count the arrows and this is the number of times you have had to halve the initial count rate of 400 counts/s.

_____ seconds [2 marks]

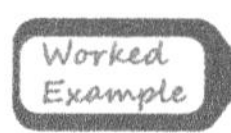

The initial activity of a radioactive isotope is 100 Bq and the half-life is 10 years. Determine how long it will it take for the activity to reduce to approximately 6 Bq.

$100 \rightarrow 50 \rightarrow 25 \rightarrow 12.5 \rightarrow 6.25$.

We have halved 100 four times here.

That means 4 half-lives will have passed. Four half-lives is 4×10 years = 40 years.

3. A Geiger–Muller tube records the number of radioactive decays per second.

Describe how you can be sure you are **only** counting the decays of the sample and not of other materials.

____

____ [2 marks]

4. **Maths**

The count rate of a radioactive source is recorded every minute. The background count is subtracted from each reading.

The table shows the results:

Time (minutes)	Count rate (counts/s)
0	200
1	175
2	146
3	120
4	96
5	83
6	68
7	57
8	48
9	38
10	33
11	30

a Plot the results from the table on the graph.  [2 marks]

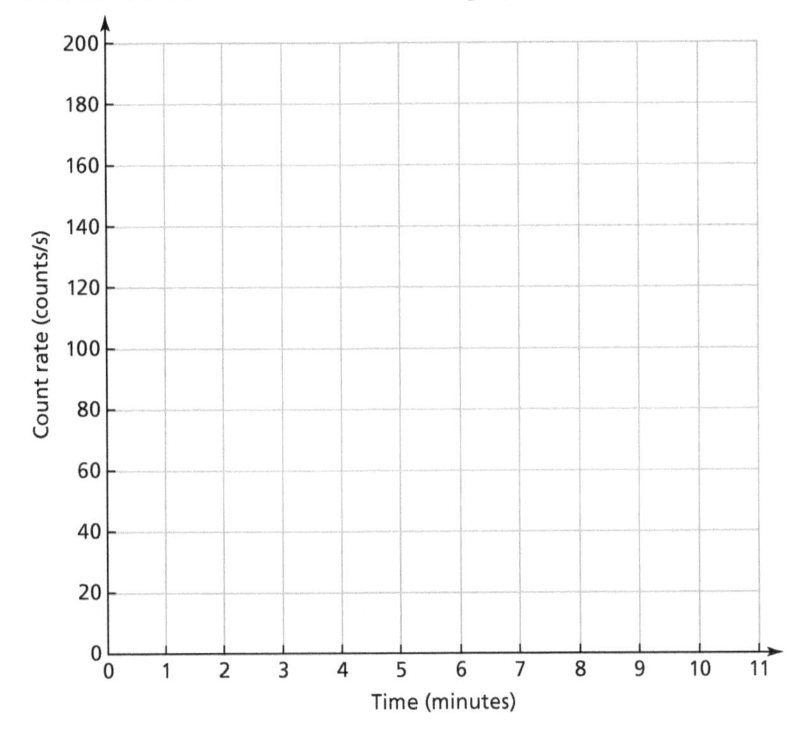

b Draw a line of best fit. [1 mark]

c Determine the time taken for the count rate to decrease from 200 to 100.

_______ minutes [1 mark]

d Determine the time taken for the count rate to decrease from 100 to 50.

_______________________________________ minutes [1 mark]

e Estimate how long it would take for the count-rate to reduce to approximately 6 counts/s.

_____ minutes [3 marks]

5. The half-life of a radioactive substance is 6 hours.

Maths Calculate how long it will take a sample of mass 0.64 g to decay until there is only 0.08 g of the original isotope.

_____ hours [3 marks]

6. 10 g of a radioactive isotope decays until 1.25 g remains. The process takes 600 years.

Maths Calculate the half-life of the material.

_____ years [3 marks]

Radioactive contamination

1. Describe **two** precautions that someone could take to reduce the radiation hazard if they were handling:

a A source of beta radiation

_____

_____ [2 marks]

b A source of gamma radiation

_____

_____ [2 marks]

2. Explain why, outside the body, the risk of harm from alpha radiation is less than from beta or gamma radiation.

____

____ [2 marks]

3. Alexander Litvinenko was poisoned in 2006 by having radioactive polonium put in his tea which he then drank.

Was the tea contaminated or irradiated?

Explain your answer.

____

______________________________ [2 marks]

Common misconception

If an object is **contaminated** radioactive material is left in the object. The radioactive material is not in contact with the object during **irradiation** – the radiation passes through, but no matter has been transferred.

4. Food is sometimes irradiated in order to kill harmful pathogens.

Explain why it is safe to eat irradiated food, unless the food comes into contact with the radioactive source.

____

____

____

____ [4 marks]

Background radiation

1. **a** State **two** natural sources of background radiation.

____

____ [2 marks]

b State **two** man-made sources of background radiation from radioactive materials.

____

____ [2 marks]

2. Explain whether you expect levels of background radiation to be the same throughout the UK.

___

___ [2 marks]

3. Explain why background count rate should be subtracted from activity readings before calculating half-life.

___

___ [2 marks]

4. Describe how the level of background radiation can be determined in a particular place.

Include in your description how you would make the result as accurate as possible.

Analysing the question

This is an extended writing question, so you need to link your sentences together in a sensible order to get full marks. Picture yourself doing the experiment so that the instructions will be clear to someone who has not done a similar experiment before.

___

___

___

___

___ [4 marks]

Uses and hazards of nuclear radiation

1. Doctors in a hospital use a radioactive gamma source as a tracer to help diagnose some medical conditions.

a Give **two** reasons why a gamma source is used as a medical tracer, instead of alpha or beta.

___

___ [2 marks]

b The doctors have four different sources they could use, with the following half-lives.

30 s	10 minutes	6 hours	10 hours

State which would be the best choice.

Explain your answer.

____

____ [2 marks]

2. Radiotherapy is the use of a high dose of radiation to kill cancer cells. One type of radiotherapy used for tumours deep within the body involves several beams of gamma rays passing through the body from different directions. The beams are targeted on the cancerous tumour.

Common misconceptions

Gamma is often thought not to be very ionising as it is the least ionising of alpha, beta and gamma. However, remember it is the **most** ionising of all of the electromagnetic spectrum of radiation – more so than X-rays and UV rays. It is very highly ionising, just less so than alpha and beta.

a Explain why **gamma** radiation has been chosen instead of alpha or beta.

____

______________________________________ [2 marks]

b Suggest why using **multiple** beams instead of one particularly intense beam is a benefit to the patient.

____

____ [2 marks]

3. The radiation dose from living near particularly radioactive rocks in Cornwall can be as much as 7.8 mSv/year, which is more than three times the national average dose received from the ground. 50 mSv/year is the recommended limit.

Command term

Evaluate means to consider evidence for and against. Here, you are asked to evaluate the risks of living in Cornwall so you should list some reasons for and against being concerned.

Evaluate whether those living in Cornwall should be concerned.

____

____

____ [2 marks]

Nuclear fission

1. Complete the sentence.

Nuclear fission occurs when a large and unstable nucleus absorbs a ____________ and splits into two smaller nuclei. [1 mark]

2. What form of electromagnetic radiation is emitted when an unstable nucleus undergoes fission? Tick **one** box.

Radio waves ☐ Gamma rays ☐ X-rays ☐ [1 mark]

3. What is emitted when an unstable nucleus undergoes fission? Tick **one** box.

A proton ☐ A neutron ☐ Two or three neutrons ☐ [1 mark]

4. **a** Describe how the fission process starts a chain reaction.

____

____

______________________________ [3 marks]

b Explain why a nuclear reactor needs a way of controlling the chain reaction.

____

______________________________ [2 marks]

c Control rods in a nuclear reactor allow the chain reaction to continue at a steady rate.

State the type of particles absorbed by the control rods.

______________ [1 mark]

5. The uranium fuel used in a nuclear power station contains a store of energy.

Synoptic Describe how the different stores of energy change as the chain reaction proceeds.

____

______________________________ [2 marks]

Nuclear fusion

1. Complete the sentence. Use a word from the box.

atoms	nuclei

In the process of nuclear fusion, ____________________ join to create a single, larger one. [1 mark]

2. **Synoptic**

In the Sun, fusion of hydrogen to helium occurs. This processes releases enormous amounts of energy.

Why is energy released in this nuclear fusion reaction?

Tick **one** box.

The mass gained when two nuclei join is converted to energy of radiation. ☐

Some of the mass of the nuclei is lost in the reaction and converted to energy of radiation. ☐ [1 mark]

3. Give **one** similarity and **one** difference between nuclear fission and nuclear fusion.

Similarity: ____

Difference: ____ [2 marks]

Scalars and vectors

1. The diagram shows towns A and B, joined by a road. The distance measured in a straight line from town A to town B is 10 km. The distance along the road from A to B is 26 km.

N

Town **A**

Town **B**

10 km

Which statement correctly describes the displacement of town B from town A? Tick **one** box.

26 km ☐

10 km due east ☐

10 km due north ☐

10 km ☐

[1 mark]

2. Which two quantities are scalar quantities? Tick **two** boxes.

Mass ☐ Displacement ☐ Time ☐ Force ☐

[2 marks]

3. Complete these sentences about vectors and scalars. Use answers from the box.

size	displacement	distance
direction	scalar	vector

A vector has both size and ____________. A scalar has ____________ but no direction. ____________ is measured in a straight line from a start point to a finish point. Distance is a ____________ quantity.

[4 marks]

4. The diagram shows the path from X to Y, with three arrows representing displacement vectors.

N

diagram scale

10 m

X

Y

What is the correct description of the three vectors that describe the path from X to Y? Tick **one** box.

20 m east, 20 m north, 20 m east ☐

10 m west, 10 m north, 20 m west ☐

10 m east, 10 m north, 20 m east ☐

10 m east, 10 m north, 10 m east ☐

[1 mark]

Speed and velocity

1. A car completes a journey of 100 km in 2.0 hours.

Maths

Calculate the average speed of the car in km/h.

Average speed = ______ km/h [2 marks]

Maths
You need to learn the equation that links distance travelled and average speed – it may **not** be given to you in a question. You also need to be able to rearrange equations.

Worked Example

A student walks 1.1 km to school in a time of 15 minutes.

Calculate her average speed.

Rearranging $s = vt$ to make speed, v, the subject gives $v = \frac{s}{t}$

$= \frac{1100 \text{ m}}{15 \times 60 \text{ s}}$

$= 1.2$ m/s

2. An aircraft travels a distance of 50 km in a straight line due north in 250 s.

Maths

Calculate the aircraft's average velocity in m/s.

Velocity = ______ m/s

Direction ______ [2 marks]

Remember
It is useful to know these typical speed estimates so you can check that your answers to velocity calculation questions are realistic: walking ≈ 1.5 m/s; jogging ≈ 3 m/s; cycling ≈ 6 m/s; car ≈ 20 m/s; aircraft ≈ 200 m/s.

3. What would be a suitable estimate for how long it takes an average runner to cover a distance of 1500 m? Tick **one** box.

Maths

5 s ☐ 50 s ☐

500 s ☐ 5000 s ☐ [1 mark]

Maths
To round a number to three significant figures, look at the fourth digit. Round up if the digit is 5 or more, and round down if the digit is 4 or less.

For example, 1.620 67 rounds to 1.62.

4. In athletics, the 100 m race is run on a straight track. A world-class sprinter wins the race in 9.90 s.

Maths

a At the finish line, what is his displacement?

Displacement = ______ m [1 mark]

b What is his average velocity?

Give your answer to three significant figures.

Average velocity = ______ m/s [3 marks]

5. The average speed of sound in air is 330 m/s.

Maths

Calculate how much time it takes for the sound of a clap of thunder from a storm to be heard 1 km away.

Time = ______ s [2 marks]

6. Maths

a A triathlete completes an iron man race which involves a swim, a cycle ride and a run. His times for the three stages of the race are shown in the table.

Complete the table to show his average speed for each of the three stages of the race.

Remember

To get the speed in m/s, you need to convert distances in km to m, and times in minutes to seconds.

Stage	Time (minutes)	Average speed (m/s)
1900 m swim	30	
90 km cycle	150	
21.1 km run	100	

[3 marks]

b Calculate the total distance he travels.

Total distance = __________ m [1 mark]

c Calculate his average speed for the whole race.

Average speed = ______ m/s [2 marks]

7. Here are different distance–time graphs for three different cars.

Remember

The gradient of an object's distance–time graph is equal to the speed of the object.

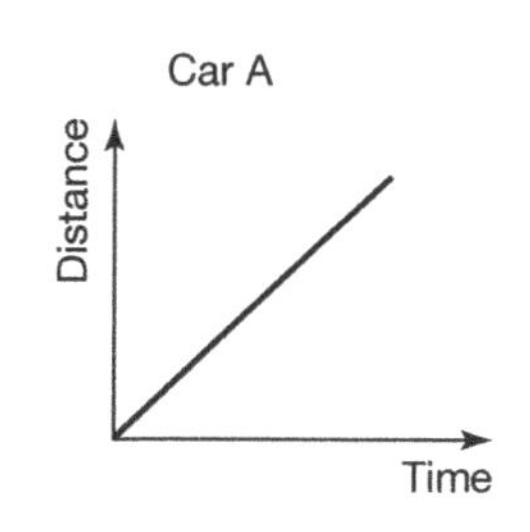

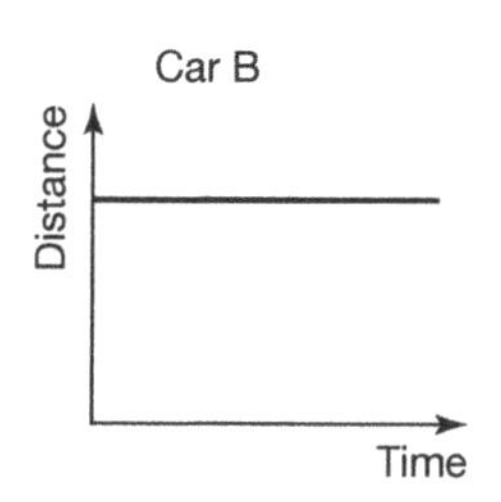

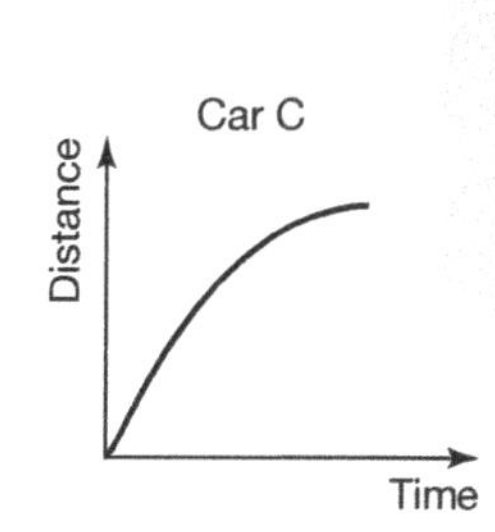

Describe the motion of each car. Use answers from the box.

getting faster	getting slower	not moving	steady speed

Car A: ______________________

Car B: ______________________

Car C: ______________________ [3 marks]

8. Maths

Here is a distance–time graph for part of a car's journey on a test track.

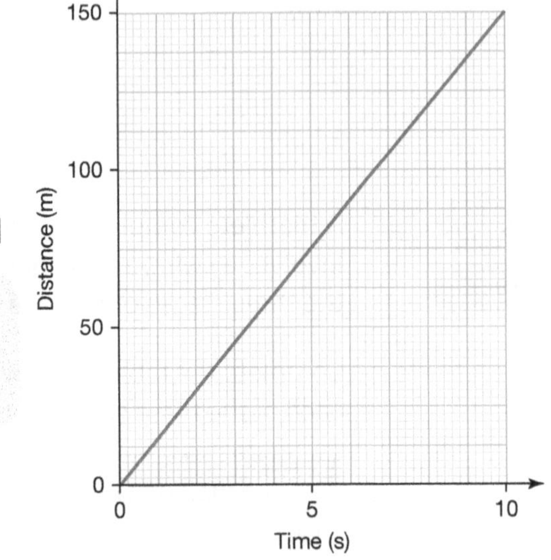

a Determine the car's speed from the graph.

Speed = ______ m/s [2 marks]

Command word

When the question asks you to **determine**, you must use the data that is given to get your answer.

b What is the correct description of the car's motion? Tick **one** box.

Getting faster ☐ Steady speed ☐ Getting slower ☐ [1 mark]

9. Higher Tier only Maths

Here is a distance–time graph for a football falling vertically.

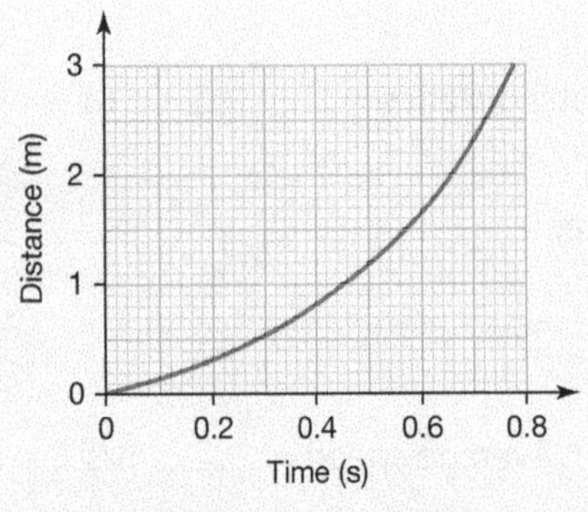

Use the graph to determine the speed of the football when it has a fallen a distance of 1.5 m.

Show clearly how you work out your answer.

Speed = ______ m/s [3 marks]

Remember

For the Higher Tier exam, you need to know that if an object's speed is changing, its speed at a certain time can be found by drawing a tangent and measuring the gradient of the tangent at that time.

Command words

The question asks you to show **how** you work out your answer from the graph. This means you **must** draw a line on the graph to show how you worked out your answer.

Acceleration

1. Calculate the acceleration of a car if its velocity increases from 2.0 m/s to 6.4 m/s in 4.0 s.

Maths

Acceleration = ______ m/s^2 [2 marks]

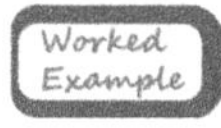

A lorry is travelling at a steady speed of 18 m/s. The driver then accelerates the lorry for 5.0 s until it reaches a speed of 23 m/s.

Calculate the lorry's acceleration.

$$\text{acceleration} = \frac{\text{change in velocity}}{\text{time taken}}$$

$$= \frac{23\text{ m/s} - 18\text{ m/s}}{5\text{ s}} = 1.0\text{ m/s}^2$$

Maths

You need to learn the equation for acceleration – it **may** not be given to you in a question.

2. A Formula 1 racing car can accelerate from a stationary position to 25 m/s in 2.1 s. Calculate the car's acceleration.

Maths

Give your answer to 2 significant figures.

Acceleration = ______ m/s^2 [3 marks]

3. A cyclist is travelling along a straight road at a steady speed of 6.5 m/s. He pedals faster for 5.0 s and reaches a speed of 7.7 m/s.

Maths

Calculate his acceleration.

Give a unit with your answer.

Acceleration = ______ Unit ______ [3 marks]

4. This is a velocity–time graph for a car on a test track.

Maths

Describe the motion of the car between the times given. Use answers from the box.

constant velocity	stationary	deceleration	acceleration

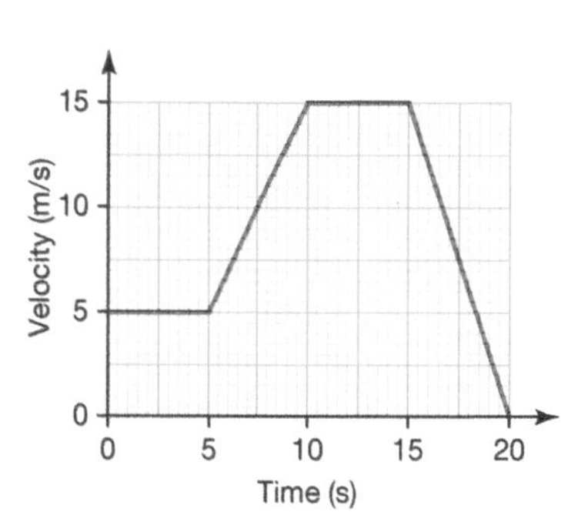

0 to 5 seconds: ______________________

5 to 10 seconds: ______________________

10 to 15 seconds: ______________________

15 to 20 seconds: ______________________ [4 marks]

5. **Maths**

The velocity–time graph for a sprinter during the first 2 seconds of a 100 m race is given below.

Calculate the sprinter's acceleration during the first 2 seconds of her race.

> **Remember**
> You need to know that the **gradient** of an object's velocity–time graph is equal to its **acceleration**.

Acceleration = ______ m/s^2 [2 marks]

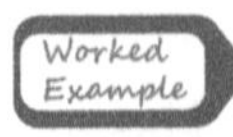

Use this velocity–time graph to calculate the sprinter's displacement during the first 2 s of the race.

Displacement = area of triangle formed by the line and the time axis up to the time of 2 s.

The area of a triangle is given by:

$$\text{area} = \frac{1}{2} \times \text{base} \times \text{height}$$

Displacement of the sprinter $= \frac{1}{2} \times 2\ \text{s} \times 6\ \text{m/s} = 6\ \text{m}$

> **Remember**
> For the **Higher tier** exam you need to know that the **area** enclosed by the line and the time axis gives the object's **displacement**.

6. **Maths**

This is the velocity–time graph for a car on a straight test track.

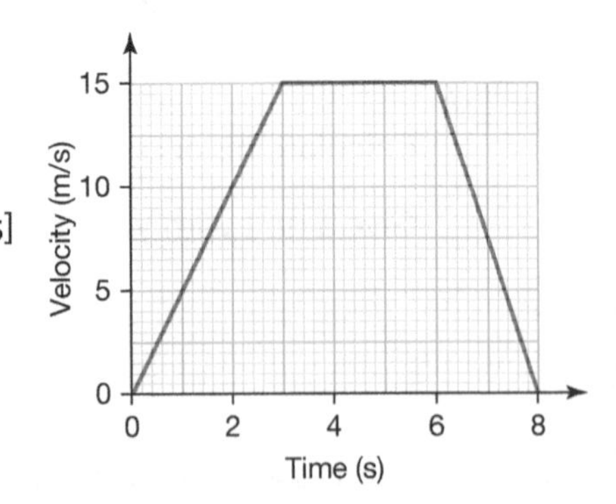

a For the first 3 s the car has uniform acceleration.

Calculate this uniform acceleration.

Acceleration = ______ m/s^2 [2 marks]

b Between 6 and 8 s, the car has uniform deceleration.

Calculate this deceleration.

Deceleration = ______ m/s^2 [2 marks]

Higher Tier only

c Calculate the car's displacement during the 8 s of its motion.

Displacement = ______ m [2 marks]

7. Maths Higher Tier only

This graph shows how the velocity of a sky diver changes from the moment she jumps out of the aircraft until her parachute opens.

Use the graph to estimate the distance she falls before opening her parachute.

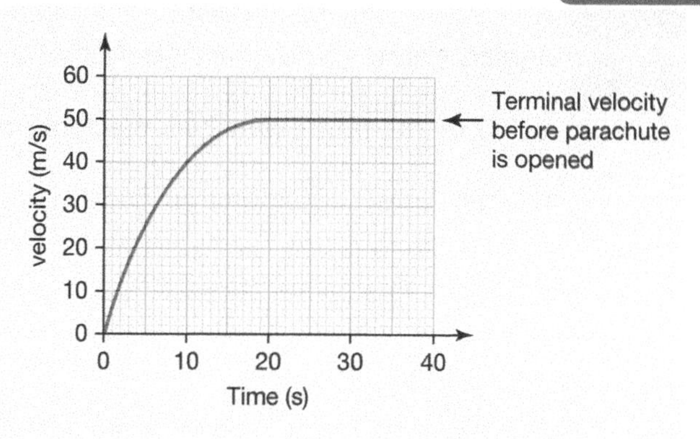

Distance fallen = ______ m [3 marks]

Equation for uniform acceleration

1. Maths

A train, initially stationary at the platform, accelerates with an acceleration of 0.50 m/s^2.

Calculate the train's velocity when it has travelled a distance of 80 m.

Use the correct equation from the Physics Equation Sheet.

Maths

This equation is on the Physics Equation Sheet. You should be able to select and apply the equation – it may **not** be given in the question.

Train's velocity = ______ m/s [2 marks]

Worked Example

A sprinter accelerates from the start over the first 20 m with an acceleration of 2.5 m/s^2.

Calculate his velocity as he passes the 20 m mark.

First match the data to the equation symbols: $u = 0$, $a = 2.5$, $s = 20$.

Now substitute the data into the equation:

$v^2 - u^2 = 2as$

$v^2 - 0 = 2 \times 2.5 \times 20$

Which gives $v^2 = 100$ so his velocity at the 20 m mark is 10 m/s.

2. Maths

A falling object on the Moon accelerates to the ground with an acceleration of 1.6 m/s^2. An astronaut on the Moon drops an object from a height of 1.3 m.

Calculate the object's velocity when it hits the ground.

Use the correct equation from the Physics Equation Sheet.

Object's velocity = ______ m/s [2 marks]

3. Maths

An electric tram moving with a velocity of 5 m/s comes to a 50 m long straight section of track and accelerates with an acceleration of 0.30 m/s^2.

Calculate the tram's velocity when it has travelled the 50m along this straight section of track.

Use the correct equation from the Physics Equation Sheet.

Tram's velocity = ______ m/s [2 marks]

Forces

1. Complete the sentences about forces. Use answers from the box.

friction	air resistance	gravitational
tension	normal contact	

Remember

'Normal' means at right angles. A normal contact force is a force at right angles to the surface where two objects are in contact.

The force in a rope being used in a tug-of-war is a ______________________ force.

The force exerted on the tyres of a bicycle when the cyclist is travelling along the road is a

______________________ force.

The upward force exerted by the ground on a person standing is a

______________________ force.

The ______________________ force is an example of a non-contact force. [4 marks]

2. The diagram shows a box on the floor. The box is being dragged to the right by a person pulling the rope.

Add force vector arrows to represent the following forces. Label them, *A*, *B*, *C* and *D*, where force *A* (force of gravity) = 50 N; force *B* (normal contact force) = 50 N; force *C* (tension in the rope) = 25 N and force *D* (friction) = 25 N [4 marks]

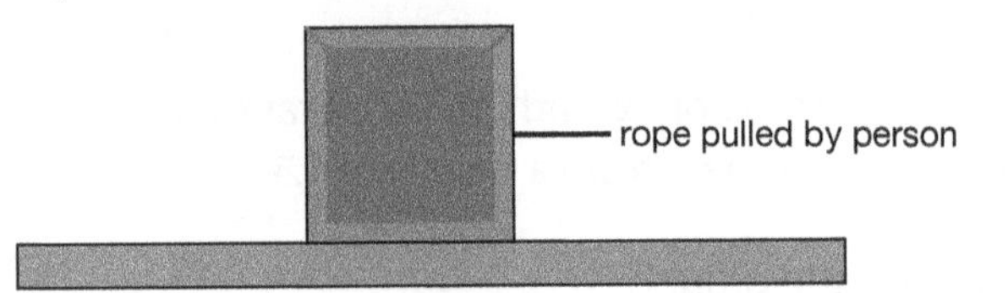

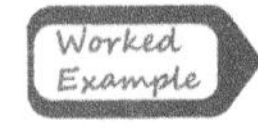

The diagram shows a book on a table.

Draw arrows to represent the forces acting on the book.

Label the arrows with the names of the forces.

The two forces acting on the book in the diagram are the force of gravity exerted by the Earth and the normal contact force exerted by the table.

Since the book is not moving these two forces are the same size so their arrows must be drawn the same size.

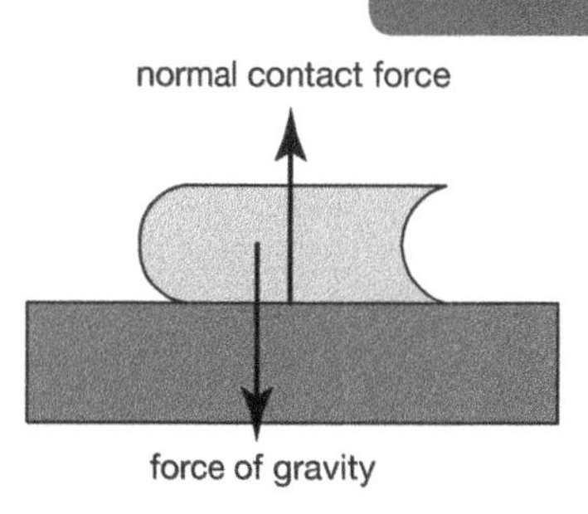

3. Identify the forces, *A*, *B*, *C*, *D*, *E*, *F* and *G* acting on the van in this diagram.

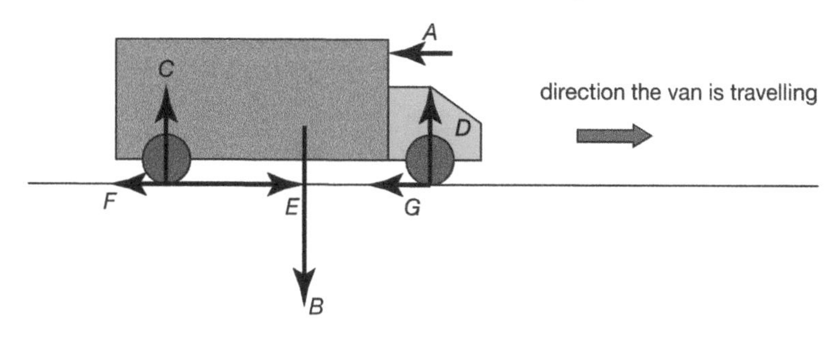

A: ______________________ *B*: ______________________

C: ______________________ *D*: ______________________

E: ______________________ *F*: ______________________

G: ______________________

[7 marks]

Moment of a force

1. Maths

When a person exerts a force on the door handle in the diagram the door rotates about the hinge and opens.

Amy pulls on the door with a force of 10 N. The distance from the handle to the hinge is 0.80 m.

Calculate the size of the clockwise moment acting on the door.

Give a unit with your answer.

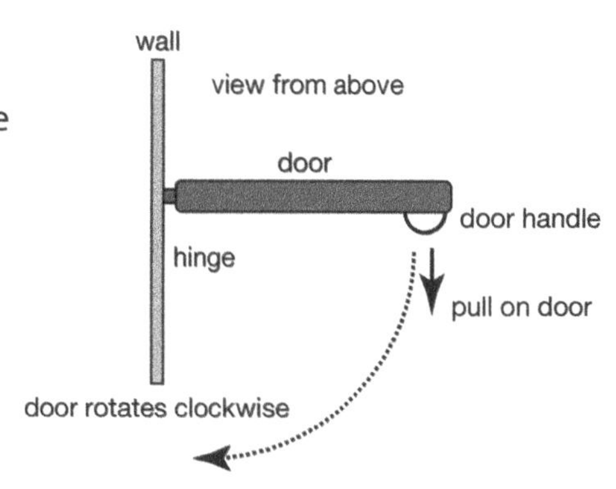

Moment ______ Unit ______ [3 marks]

Maths

You need to know the equation for the moment of a force – it **may** not be given to you in a question.

2. Maths

A student uses the apparatus in the diagram shown in the worked example below to find the weight of a pebble. She positions the 1.0 N weight 21 cm from the pivot. She finds that the 50 cm rule is balanced if she puts the pebble 16 cm from the pivot.

Calculate the weight of the pebble. Give your answer to 2 significant figures.

Weight of pebble = ______ N [3 marks]

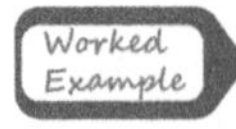

A student uses the apparatus shown in the diagram to find the weight of object X. The pivot is positioned at the centre of the 50 cm rule. The student puts the 1.0 N weight 21 cm from the pivot. She finds that the rule balances if she puts X 10 cm from the pivot.

Calculate the weight of X.

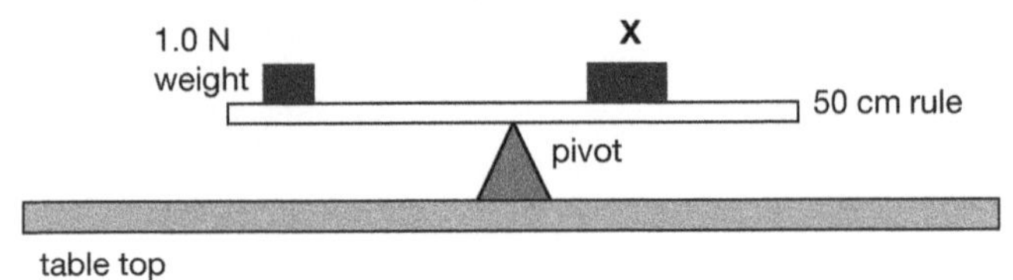

The 1.0 N weight is 21 cm from the pivot. X is 10 cm from the pivot.

Anticlockwise moment = clockwise moment

$1.0 \text{ N} \times 21 \text{ cm} = X \times 10 \text{ cm}$

which gives $X = 21 \div 10 = 2.1$ N

3. Maths

a Another student uses the apparatus in the same diagram to balance the half-metre rule with a 1.0 N weight on the left and a 2.8 N weight on the right-hand side. The 1.0 N weight is placed 20 cm from the pivot.

Calculate the anticlockwise moment of the 1.0 N weight. Give your answer in N cm.

Anticlockwise moment = ______ N cm [2 marks]

b Calculate how far from the pivot the 2.8 N weight must be placed if the rule is to balance.

Give your answer to 2 significant figures.

Distance from pivot = ______ cm [3 marks]

c The zero of the half meter rule is on the left of the diagram.

Use your answer to part **b** to find which cm mark on the rule the 2.8 N weight must be closest to in order to balance the rule.

Mark on rule is ______ cm [2 marks]

Levers and gears

1. Maths

A gardener wants to move a large boulder. The weight of boulder is 800 N. To lift the boulder the gardener would have to exert a force of 800 N.

He decides to use a plank of wood as a lever to make the job easier.

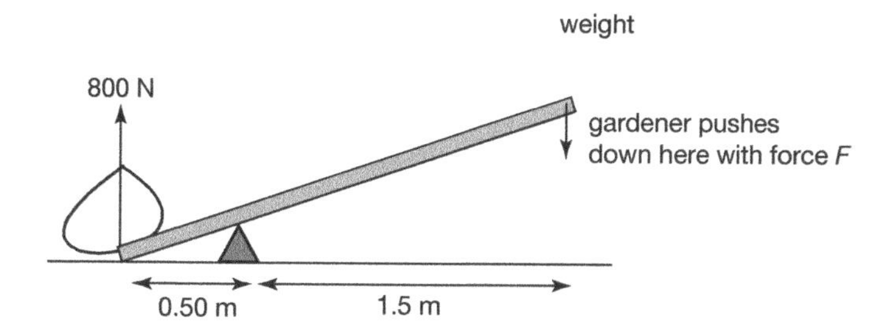

a To lift the boulder, the lower end of the plank has to exert an upward force of 800 N.

Calculate the moment about the pivot needed to lift the boulder.

Give the unit with your answer.

Moment = ______ Unit ______ [3 marks]

b The force F exerted by the gardener creates the moment needed to lift the boulder.

Calculate the value of force F.

F = ______ N [2 marks]

2. On a bicycle, front and back gears along with the chain are used to transmit the rotational effect of the force applied to the pedals to the back wheel.

Use answers from the box to complete the sentences.

large	small

When a cyclist is travelling uphill the front gear should be ________________ and the back gear should be large. When a cyclist is travelling downhill the front gear should be ________________ and the back gear should be ________________. [3 marks]

3. Maths

To move the soil in the diagram, the gardener has to lift the handles of the wheelbarrow so that the back legs are just off the ground. The weight of the wheelbarrow and soil is 200 N.

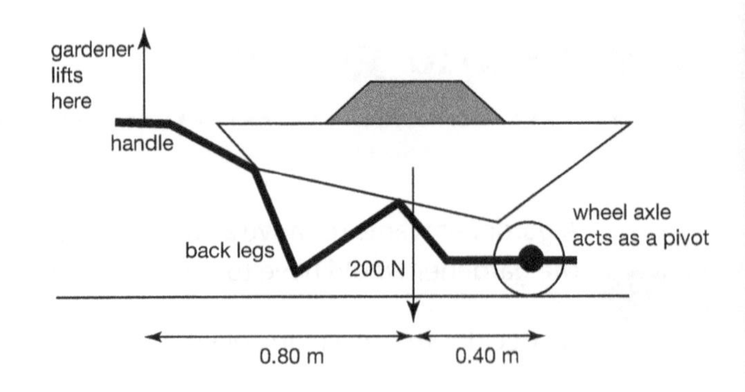

a Calculate the moment of the weight of the wheelbarrow and soil about the pivot.

Moment = ______ N m [2 marks]

b Is the moment calculated in part **a** clockwise or anticlockwise?

______________________ [1 mark]

c To just lift the wheelbarrow, the gardener needs to create a moment equal but opposite to that calculated in part **a**.

Calculate the size of the force that the gardener exerts on the handles.

Force on handle = ______ N [2 marks]

Pressure in a fluid

1. Maths

The base of a kettle has an area of 0.018 m^2. The water in the kettle exerts a force of 10 N on the base of the kettle.

Calculate the pressure exerted by the water on the base of the kettle.

Give your answer to three significant figures.

> **Maths**
> You need to know the equations for pressure on a surface and pressure due to a column of liquid – they may **not** be given in a question.

Pressure exerted by the water = ______ Pa [3 marks]

2. Maths

The diagram shows a treasure chest lying at the bottom of the sea at a depth of 100 m.

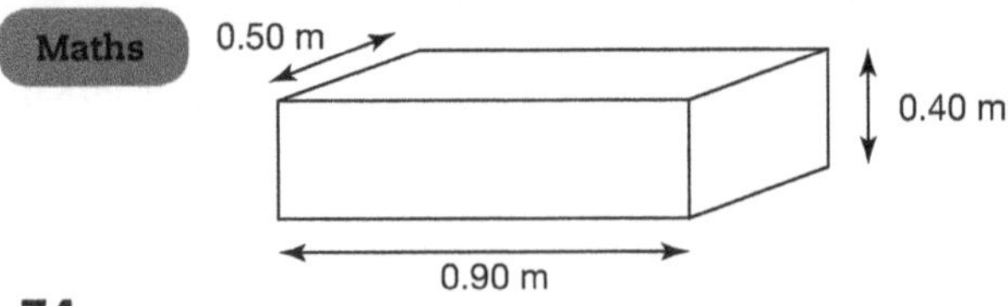

a Calculate the area of the lid of the box.

Area of box lid = ______ m^2 [1 mark]

b The pressure of water on the chest is 1 100 000 Pa.

Calculate the force exerted on the lid of the chest.

Give the unit with your answer.

Force exerted on lid = ______ Unit ______ [3 marks]

3. Maths

The surface area of an average adult human is 1.7 m^2. At sea level the pressure due to the atmosphere is about 100 000 Pa.

Calculate the total force exerted by the atmosphere on the human body at sea level.

Maths

You will need to rearrange the pressure equation for this calculation. This means putting force on its own, on one side of the equals sign.

The equation is in the form $a = b/c$. To find b, multiply both sides of the equation by c.

Force exerted = ______ N [2 marks]

4. Maths Higher Tier only

A scuba diver is 15.0 m below the surface of the ocean.

Density of seawater = 1030 kg/m^3

Gravitational field strength = 9.8 N/kg.

a Calculate the pressure due to the sea water on the diver.

Use the correct equation from the Physics Equation Sheet.

Pressure due to the water = ______ Pa [2 marks]

b The total pressure at a depth of 15.0 m is equal to the pressure due to the weight of the sea water and the pressure due to the atmosphere. Atmospheric pressure at sea level is 101 000 Pa.

Calculate the total pressure at a depth of 15.0 m below the sea.

Total pressure = ______ Pa [1 mark]

c The greatest ocean depth reached by a manned submarine is 10 900 m.

Calculate the pressure exerted at this depth due to the sea water.

Pressure due to the sea water = ________________ Pa [2 marks]

d How many times bigger is the pressure due to the sea water at a depth of 10 900 m compared with atmospheric pressure at sea level?

Number of times greater = ______ [1 mark]

Atmospheric pressure

1. The graph shows how atmospheric pressure decreases as height above sea level increases.

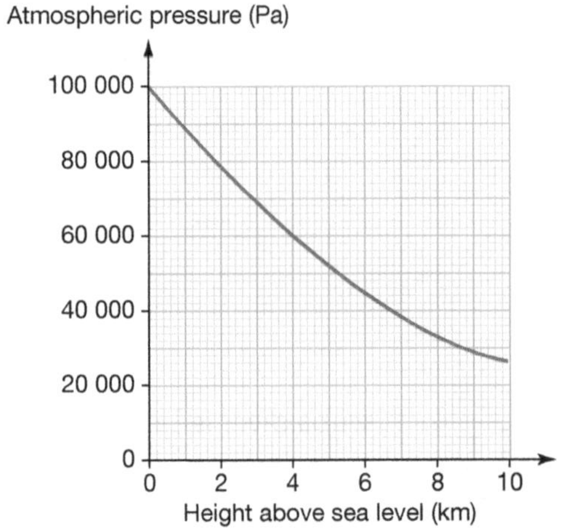

Atmospheric pressure (Pa)

100 000
80 000
60 000
40 000
20 000
0

0 2 4 6 8 10

Height above sea level (km)

Maths **a** Use the graph to calculate the force exerted by the atmosphere on a person at the top of Mount Everest, approximately 9000 m above sea level.

Take the average surface area of an adult human to be 1.7 m^2.

Force = ________________ N [3 marks]

b Explain how air exerts a pressure on surfaces.

____

____ [2 marks]

c Molecules in the air at sea level are closer together than at the top of Mount Everest.

Explain why air with more molecules in each cubic metre creates a bigger pressure on surfaces.

____

____ [2 marks]

d Explain why atmospheric pressure is greater at sea level than at the top of Mount Everest.

____

____

______________________________ [2 marks]

2. The thickness of the Earth's atmosphere is about 60 miles but it is often described as a thin layer of air round the Earth.

Suggest why the atmosphere is described as a thin layer.

____

______________________________ [1 mark]

3. a The diagram shows a straw in a glass of milkshake. Add an arrow to represent the direction of the force exerted by the atmosphere on the liquid.

[1 mark]

b Explain how sucking on the straw enables a person to drink the liquid.

____

____

____

______________________________ [3 marks]

Gravity and weight

1. Mass is the amount of substance in an object. Weight is the force of gravity on the object. Complete the sentences using answers from the box. Note that gravitational field strength at the Earth surface is greater than on the Moon.

increases	decreases	doesn't change

When an astronaut travels from the Earth to the Moon, his mass

______________ and his weight ______________. [2 marks]

2. *Curiosity*, NASA's Mars rover, has a mass of 900 kg.

Maths

Gravitational field strength at the Earth's surface = 9.8 N/kg

Gravitational field strength at the surface of Mars = 3.7 N/kg

Maths

The symbol for gravitational field strength is *g*. You need to know the equation for weight – it may not be given in a question.

a Calculate the weight of *Curiosity* on the Earth. Give the unit with your answer.

Weight on Earth = ______ Unit ______ [3 marks]

b Calculate the weight of *Curiosity* on Mars.

Weight on Mars = ______ Unit ______ [3 marks]

3. Weight and mass are described as being directly proportional. This relationship can be represented by the formula:

$$\text{weight} \propto \text{mass}$$

Explain what is meant by describing weight and mass as being **directly proportional**.

____

____ [2 marks]

4. The diagram shows a block of wood.

Draw a cross on the diagram to mark the position of the centre of mass of the block.

Draw an arrow to represent the weight of the block. [2 marks]

5. a Name the apparatus used to measure the weight of an object.

______________________ [1 mark]

Maths

b A student measures the weight of an object and obtains a value of 2.7 N. Calculate the mass of the object.

Gravitational field strength = 9.8 N/kg.

Mass of object = ______ N [2 marks]

6. Maths

The fuel in a car's full tank of petrol has a mass of 4.3 kg. After a journey, the mass of fuel remaining in the tank is 2.6 kg.

Calculate the reduction in the car's weight as a result of the journey.

Give your answer to two significant figures.

Gravitational field strength = 9.8 N/kg.

Reduction in car's weight = ______ N [4 marks]

Resultant forces and Newton's 1st law

1. Maths

The diagram shows a child's trolley and the forces acting on it. One child exerts a force of 5 N in one direction and another child exerts a force of 11 N in the opposite direction.

Calculate the resultant force and state its direction.

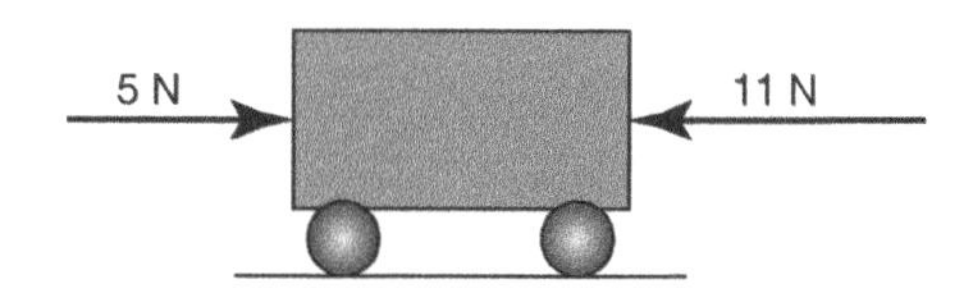

Resultant force = ______ N

Direction ______________ [2 marks]

2. The diagram shows a crate being dragged along the ground by a person pulling a rope. The arrows show the pull on the rope and friction between the crate and the floor.

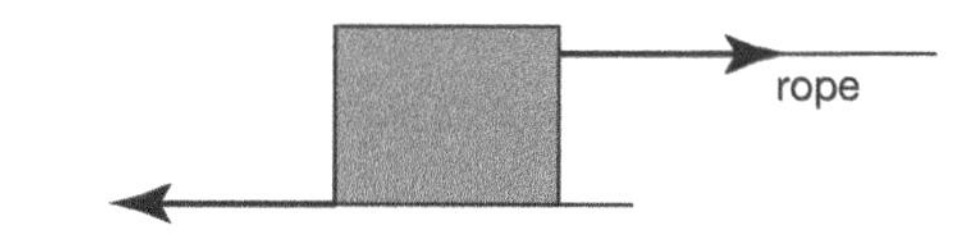

a State Newton's first law as applied to a moving object.

____

____ [2 marks]

b The crate is moving at a steady speed and the size of the pulling force is 50 N.

What is the resultant force on the crate?

Resultant force = ______ N [1 mark]

3. **Higher Tier only**

A sky diver with her parachute open is falling vertically at a steady speed towards the ground.

Draw a free body diagram in the box to represent the forces on the sky diver.

Label the diagram with the names of the forces acting on her. [3 marks]

Remember

When drawing arrows to represent vectors, e.g. forces, the **length** of the arrow represents the **size** of the vector.

4. **Maths** **Higher Tier only**

The arrow in the diagram represents the tension force in the string in a kite that is flying.

20 N

a Complete the diagram to show the horizontal and vertical components of the tension force. [2 marks]

b The arrow has been drawn to scale.

Determine the size of the vertical and horizontal components of the tension force.

Horizontal component = ______ N

Vertical component = ______ N [2 marks]

Forces and acceleration

1. **Practical**

The diagram shows a glider on an air track. Two light gates are placed above the track so that the card attached to the glider passes through the gates as the glider moves.

A student uses the apparatus to investigate acceleration and resultant force. The glider is accelerated by a weight that it is attached to it by string. The glider is released at the left end of the air track and passes through both light gates.

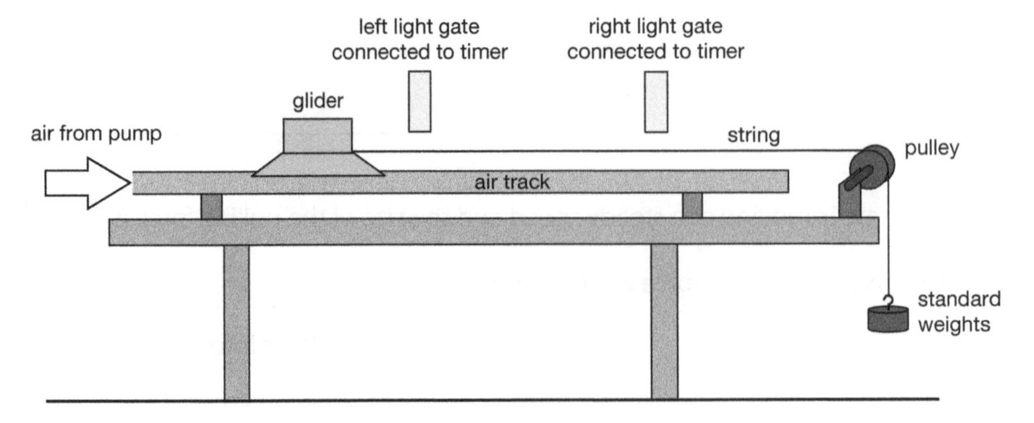

a What is the advantage of using the air track for motion experiments?

______________________________ [1 mark]

b What force on the glider is reduced by its streamlined shape?

______________________________ [1 mark]

c The card on top of the glider is 0.10 m wide. When the first weight is attached to the string the card passes through the light gate on the left in 0.77 s.

Use the equation

$$\text{speed} = \frac{\text{distance travelled}}{\text{time}}$$

to calculate the glider's velocity as it passes through the first gate.

Initial velocity of the glider = ______ m/s [2 marks]

d The glider passes through the light gate on the right in a time of 0.30 s. Calculate the glider's velocity as it passes through the gate.

Give your answer to 2 significant figures.

Final velocity of the glider = ______ m/s [3 marks]

e The distance between the light gates is 0.50 m.

Use the following equation to calculate the acceleration of the glider:

$$(\text{final velocity})^2 - (\text{initial velocity})^2 = 2 \times \text{acceleration} \times \text{distance}$$

Remember

This equation is on the Physics Equation Sheet. You should be able to select and apply the equation – it may **not** be given in the question.

Glider's acceleration = __________ m/s^2 [2 marks]

f The student repeats the experiment with different weights attached to the string.

The student assumes that the force accelerating the glider is equal to the

Practical

Anomalous values, also known as **outliers**, are values which you judge not to be part of the variation due to random uncertainty.

weight attached to the string. The graph shows how he plots his results.

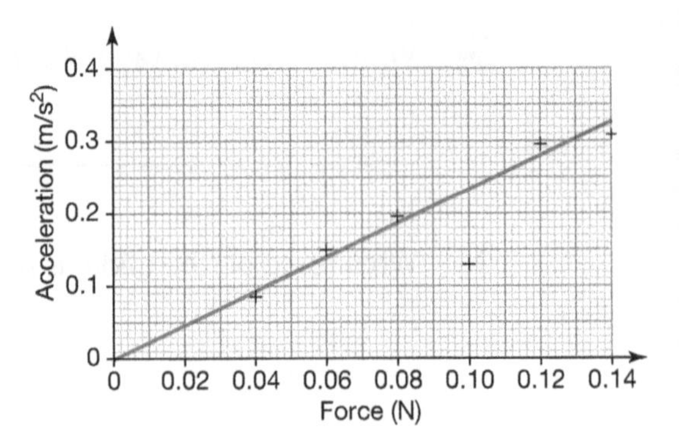

The student decides that one of the data sets is anomalous. He ignores the anomalous point and draws the line of best fit through the other points.

What should the student have done instead of ignoring the anomalous point?

____

____ [2 marks]

g What could the student conclude from the best fit line that he has drawn?

Explain your answer.

____

____ [2 marks]

2. The engine driving force on a lorry is 4000 N. The total drag forces add up to 1200 N. The mass of the lorry is 2000 kg.

Maths

a Calculate the resultant force on the lorry.

Resultant force = ______ N [1 mark]

b Use the equation $a = F \div m$ to calculate the lorry's acceleration.

Give the unit with your answer.

Lorry's acceleration = ______ Unit ______ [2 marks]

Worked Example

The engine driving force on a car is 2000 N. The total drag forces add up to 700 N. The mass of the car is 900 kg.

Calculate the car's acceleration.

The resultant force on the car = 2000 − 700 = 1300 N.

$F = ma$

so $a = F \div m$

$= 1300 \div 900 = 1.4\ m/s^2$

Maths

You need to learn the equation that links resultant force, mass and acceleration – it may **not** be given in the question.

3. Maths

NASA's Falcon-9 rocket is launched to take a satellite into orbit. At launch the thrust from the rocket's engines is 8 200 000 N. The air resistance acting on the rocket is 600 000 N. The weight of the rocket is 5 500 000 N.

Command words

When a question asks you to state, it is asking you to write down simple information – there is no need to give reasons.

a State whether the thrust, the air resistance and the weight act upwards or downwards on the rocket.

Thrust: ______________________

Air resistance: ______________________

Weight: ______________________ [3 marks]

b Calculate the resultant force on the rocket during the launch.

Resultant force = ______________ N [1 mark]

c Use the equation $a = F \div m$ to calculate the acceleration of the rocket given that its mass is 550 000 kg.

Acceleration = ______ m/s^2 [2 marks]

4. Maths

a The diagram shows an apple falling from a tree. The forces on the apple as it falls are its weight and air resistance. The apple's weight is bigger than the air resistance.

motion of apple

Add arrows to the diagram to represent these two forces.

Label each arrow with the name of the force. [3 marks]

b The mass of the apple is 100 g. Give the apple's mass in kg.

Mass of apple = ______ kg [1 mark]

c Calculate the weight of the apple.

Gravitational field strength = 9.8 N/kg.

Weight of apple = ______ N [1 mark]

d If the air resistance is 0.30 N, calculate the resultant force on the apple.

Resultant force = ______ N [1 mark].

e Calculate the acceleration of the apple.

Acceleration = ______ m/s^2 [2 marks]

Terminal velocity

1. The diagram on the right shows a steel ball released at the surface of the oil in a glass tube. The graph shows how the velocity of the ball changes as time passes.

Remember

The gradient of a velocity–time graph gives the acceleration.

a At which position is the ball's acceleration at its maximum value? Tick **one** box.

A ☐ B ☐ C ☐ [1 mark]

For parts **b**, **c** and **d**, choose your answers from this box.

not changing	increasing	zero	decreasing

b Describe the velocity of the sphere at positions A, B and C.

Position A: ______________________

Position B: ______________________

Position C: ______________________ [3 marks]

c Describe the acceleration of the sphere at positions B and C.

Position B: ______________________

Position C: ______________________ [2 marks]

d Describe the resultant force of the sphere at positions B and C.

Position B: ______________________________

Position C: ______________________________

[2 marks]

Common misconception

Sometimes students make the mistake of thinking that if the resultant force on a moving object becomes zero, it will stop. Newton's 1st law tells us that the object will actually continue with a **constant velocity**.

2. This is the velocity–time graph of a sky diver.

Describe how the velocity of the sky diver changes as time passes. Describe how the acceleration of the sky diver changes as time passes

Explain what causes the changes in the sky diver's motion. Your explanation should consider how the effect of air resistance compares in size with the sky diver's weight.

parachute opened

Velocity

Time

Command words

This question asks you to describe, then explain. Make sure you answer **both** parts of the question.

____

____

____

____

____

____

____ [6 marks]

Newton's third law

1. When two objects interact they each exert a force on the other one.

a Which is the correct description of the directions of the two forces? Tick **one** box.

Opposite direction ☐ Same direction ☐ [1 mark]

b Which is the correct description of the size of the two forces? Tick **one** box.

Same size ☐ Different sizes ☐ [1 mark]

c Which is the correct description of the type of forces acting? Tick **one** box.

Same type ☐ Different type ☐ [1 mark]

2. Complete the sentence to summarise Newton's 3rd law by filling in the letters **A** or **B**.

If body A exerts a force on body B then body _____ exerts an equal but opposite force on body _____. [1 mark]

3. The diagram shows the forces acting on a box stationary on the floor.

force **B**

force **A**

a What type of force is force A?

_______________________ [1 mark]

b What type of force is force B?

_______________________ [1 mark]

Common misconceptions

The two forces in the diagram are equal in size and acting in opposite directions and some students may mistakenly think that these are a Newton's 3rd law pair. They are **not**. The forces in a Newton's 3rd law pair act on **different** objects.

c State the type, direction and the body acted upon by the force that forms the Newton's 3rd law pair force with force A.

Type: _______________________

Direction: _______________________

Body acted upon: _______________________ [3 marks]

d State the type, direction and the body acted upon by the force that forms the Newton's 3rd law pair force with force B.

Type: _______________________

Direction: _______________________

Body acted upon: _______________________ [3 marks]

4. In a rocket engine, exhaust gases at high pressure and temperature are produced by combustion of the rocket fuel with an oxidiser in the combustion chamber.

Use Newton's 3rd law to explain how this results in the rocket being propelled upwards.

____

____

____ [2 marks]

Work done and energy transfer

1. A person is pushing a shopping trolley at a steady speed at a supermarket.

Maths

You need to learn the equation for work done – it may **not** be given in the question.

Maths **a** They push the trolley a distance of 10 m with a horizontal force of 15 N.

Calculate the work done by the force in moving the trolley 10 m.

Give a unit with your answer.

Work done = ______ Unit ______ [3 marks]

b Since the trolley is being pushed at a steady speed, the resultant force on the trolley must be zero.

Name the other horizontal force that must be acting on the trolley.

State its direction.

Name of force: ______________________

Direction of force: ______________________ [2 marks]

c Describe how the different energy stores change as the shopper is pushing the trolley.

____

____ [2 marks]

2. Maths

The diagram shows a winch dragging a 200 kg load up a slope in a loading bay. The winch exerts a force of 1000 N.

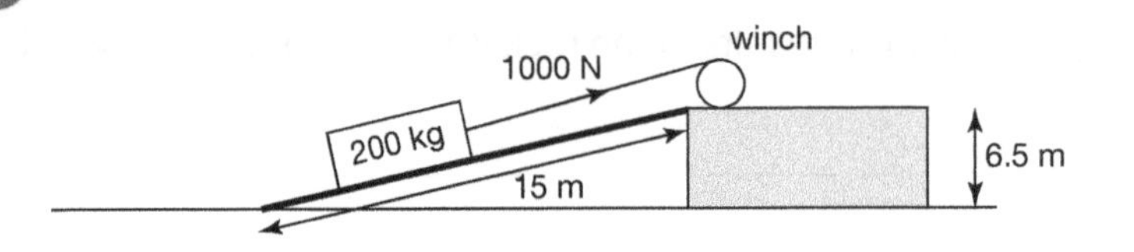

Remember

The definition of work depends on the force used and the distance moved in the direction of the force. Look carefully at the diagram to see what direction the force moves, and so which distance measurement to use.

a Use the equation $W = Fd$ to calculate how much work is done by the winch in dragging the container the 15 m up the slope.

Work done = __________ J [2 marks]

b Use the equation $E_p = mgh$ to calculate the gravitational potential energy store gained by the load when it is 6.5 m above the floor.

Gravitational field strength = 9.8 N/kg.

Gain in gravitational potential energy store = __________ J [2 marks]

c The work done by the winch is greater than the gravitational potential energy store gained by the load. Give one reason for this.

______________________________ [1 mark]

3. Maths

Remember

Power is equal to the work done per second.

a A small aircraft flies at a constant altitude at constant velocity of 100 m/s. The thrust from the engine is 800 N.

Calculate the useful work done by the engine when the aircraft travels 100 m.

Useful work done by the engine = __________ J [2 marks]

b What is the useful output power of the aircraft's engine?

Useful output power of the engine = __________ W [1 mark]

Stopping distance

1. Maths

A car is being driven at 15 m/s (about 34 mph). The mass of the car including the driver is 800 kg.

Remember

You need to know that:

stopping distance = thinking distance + braking distance

a Write down the equation which links speed, kinetic energy and mass.

______________________________ [1 mark]

Calculate the kinetic energy store of the car.

Kinetic energy store of the car = __________ J [2 marks]

b The driver sees a hazard in the road ahead. He needs to stop his car before the hazard is reached. During the time it takes time for the driver to react to the situation the car carries on moving at 15 m/s. The driver is alert and his reaction time is 0.60 s.

Use the equation

distance travelled = speed × time

to calculate the distance travelled by the car while the driver is thinking about the situation.

Thinking distance = ______ m [2 marks]

c The driver applies the brakes creating a braking force of 5000 N. For the car to stop, all its kinetic energy has to be dissipated when work is done by the brakes.

Use the equation

work done = force × distance

to calculate the distance over which the braking force must act to stop the car.

Braking distance = ______ m [2 marks]

d Calculate the car's total stopping distance.

Stopping distance = ______ m [1 mark]

2. Maths

A car is being driven at 30 m/s (about 68 mph). The mass of the car including the driver is 800 kg.

a Calculate the kinetic energy store of the car.

Kinetic energy store of the car = ____________ J [2 marks]

b Calculate the thinking distance in this situation, if the driver's reaction time is 0.60 s.

Thinking distance = ____________ m [2 marks]

c The car's braking system creates a braking force of 5000 N.

Calculate the distance over which the braking force must act to stop the car.

Braking distance = ____________ m [2 marks]

d Calculate the car's stopping distance.

Stopping distance = ____________ m [1 mark]

e Compare the stopping distance for a car travelling at 15 m/s with a car travelling at 30 m/s (use your answers to **1d** and **2d**).

Complete the sentence:

If the speed of a car is doubled then the distance that a car travels before it can stop

is ______ times bigger. [1 mark]

3. Give two factors which could increase a driver's reaction time.

____

____ [2 marks]

4. Maths

The graph shows the braking distance and thinking distance at different speeds for a typical car.

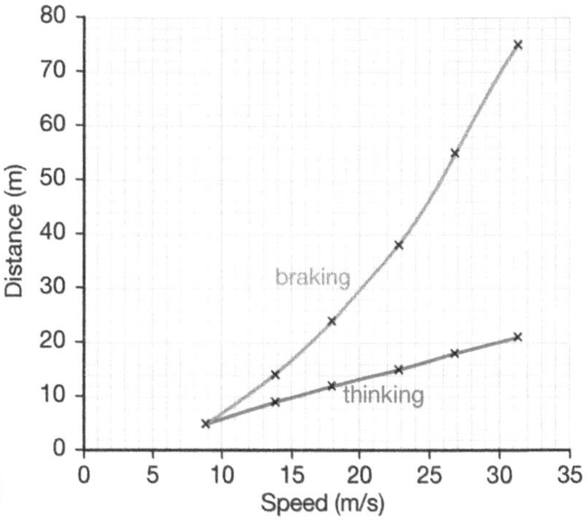

Distance (m)
80
70
60
50
40
30
20
10
0
0 5 10 15 20 25 30 35
Speed (m/s)
braking
thinking

a Use this graph to determine the stopping distance for a car travelling at 20 m/s.

Show on the graph how you obtained the answer.

Stopping distance = ______ m [2 marks]

b Give two reasons why a car travelling at 20 m/s may have a longer stopping distance than the answer obtained in part **a**.

____

______________________________ [2 marks]

Command words

Part **a** of the question asks you to show how you work out your answer from the graph. You must draw lines on the graph **as well as** showing your working.

5. **a** Describe what happens to the car's store of kinetic energy when a car driver applies the brakes, bringing the car to a stop.

____

______________________________ [1 mark]

b How are the car's brakes affected by the car's braking?

______________________________ [1 mark]

c A car is travelling at 30 mph (13 m/s). The driver does an emergency stop. His car stops in a shorter distance than that predicted by the graph.

Suggest how the brakes might be affected by this rapid deceleration.

What effect could this have on the car?

____

______________________________ [2 marks]

Command words

This question asks you to **make suggestions**. Your answer should **link** what you already know and the information in the question.

d When a car rapidly decelerates during an emergency stop, the driver's body also decelerates rapidly.

What exerts the force on the driver to cause him to decelerate?

______________________________ [1 mark]

Force and extension

1. The diagram shows a spring before and after it has been stretched.

Maths

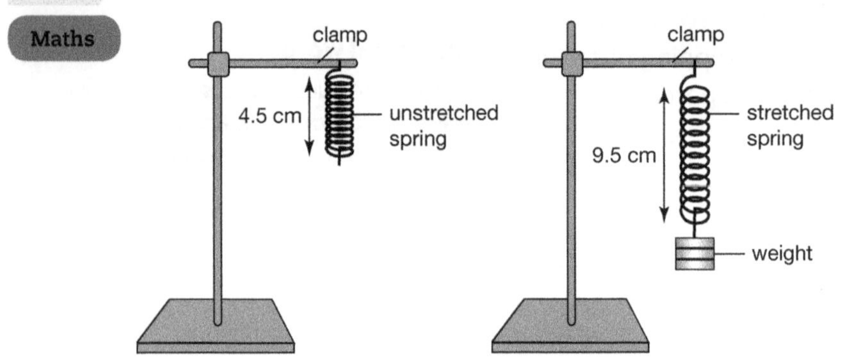

a Stretching or compressing a spring needs two forces to act on the spring.

Which two objects in the diagram are exerting forces on the spring to make it stretch?

1 ______________ **2** ______________ [2 marks]

b Calculate the extension of the spring in metres.

Extension = ______ m [1 mark]

Remember
You need to know that the extension of a spring is found by **subtracting** its unstretched length from its stretched length]

c Use the equation below to calculate the elastic potential energy E_e, stored in the stretched spring. The spring constant is 35 N/m.

Give your answer to 2 significant figures.

$$E_e = \frac{1}{2} \times \text{spring constant} \times (\text{extension})^2$$

Elastic potential energy = __________ J [3 marks]

2. Maths

A student is given four springs. Each one has a different value for its spring constant, k. So some of the springs are easy to stretch and some are harder to stretch. She adds weights to each spring to produce a stretching force, F, and measures the extension, e.

The table below shows her results.

Maths
You need to learn the equation that links the force applied to a spring to its extension – it may **not** be given to you in a question.

a Use the equation force = spring constant × extension to complete the last column in the table.

Spring number	Force F (N)	Extension e (m)	Spring constant k (N/m)
1	5.0	0.25	
2	3.0	0.10	
3	2.0	0.20	
4	4.0	0.10	

[4 marks]

b Which spring is the easiest to stretch?

Spring number ______ [1 mark]

c Which is the stiffest spring?

Spring number ______ [1 mark]

3. Practical

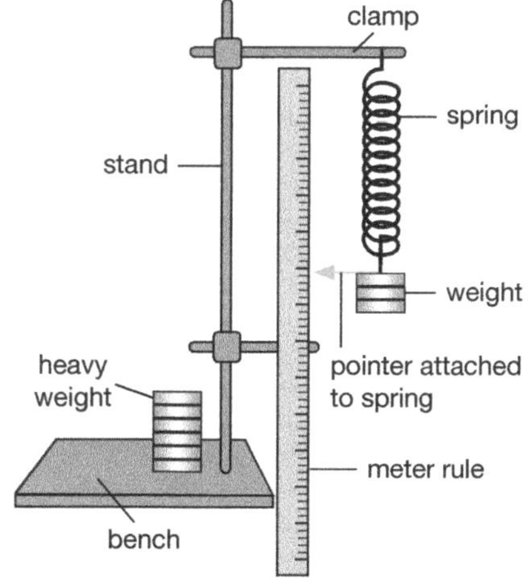

Practical
If vertical distance measurements are to be made with a clamped metre rule, first check that the rule is vertical. You can do this with a plumb line.

A student is given the apparatus shown in the diagram to investigate the relationship between extension and stretching force for a spring.

The weight attached to the spring causes a stretching force on the spring.

The metre rule is positioned so that a pointer taped to the spring points to a value on the metre rule that gives the spring's length.

The student's measurements are shown in the table on the next page.

a The unstretched length of the spring is 5.0 cm.

Complete the last column of the student's table of measurements to show the extension values of the spring in cm.

Stretching force (N)	Length of spring (cm)	Extension (cm)
0.50	7.3	
1.00	9.5	
1.50	11.8	
2.00	14.1	
2.50	16.4	
3.00	18.7	

[2 marks]

b Suggest why a heavy weight has been placed on the base of the stand.

______________________________ [1 mark]

c The student plots the measurements in the table on a graph of extension versus stretching force.

The student concludes that extension ∝ stretching force.

State what this means.

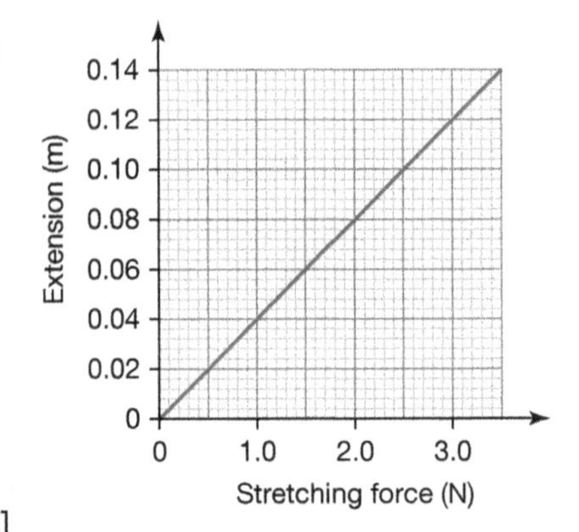

____

______________________________ [1 marks]

d Explain how the graph supports this conclusion.

____

______________________________ [2 marks]

e The student is given a stone and asked to find its weight using the apparatus and the graph above. He attaches the stone to the spring. The stone causes an extension of 11.0 cm.

Use the graph to determine the weight of the stone. Show on the graph how you obtained the answer.

Weight = ______ N [2 marks]

Momentum

1. Use the equation $p = mv$ to work out the units for momentum.

Higher Tier only

Units of momentum ____________ [1 mark]

Maths

You need to learn the equation for momentum – it may **not** be given in the question.

2. **a** Is momentum a vector or scalar quantity? Tick **one** box.

Higher Tier only

Vector ☐ Scalar ☐ [1 mark]

b Explain your answer to part **a**.

____ [2 marks]

3. A car of mass 800 kg is travelling due west on a straight section of road at a speed of 20 m/s.

Maths

Calculate the car's momentum.

Higher Tier only

Give the unit with your answer.

Momentum = ____________ due west

Unit ____________ [3 marks]

4. A 0.057 kg tennis ball is travelling at 100 km/hr.

Maths

Calculate the size of the ball's momentum.

Higher Tier only

Give your answer to 2 significant figures.

Momentum = ________ kg m/s [4 marks]

5. The motion of an 800 kg car on a straight test track is shown in the graph.

Maths

Higher Tier only

Use the diagram to calculate the size of the car's momentum at 1 s, 2 s and 3 s after the start of the car's motion.

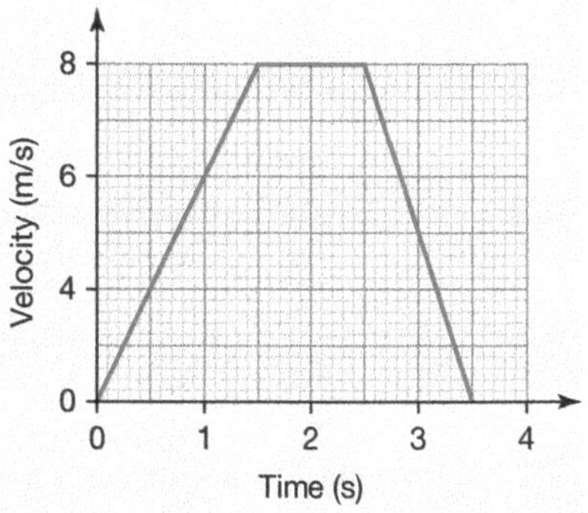

Show clearly how you work out your answer.

Momentum at 1 s: ______ kg m/s

Momentum at 2 s: ______ kg m/s

Momentum at 3 s: ______ kg m/s [3 marks]

6. A car of mass 800 kg has momentum equal to 1200 kg m/s.

What is the velocity of the car?

Car's velocity = ______ m/s [2 marks]

Conservation of momentum

1. State what is meant by 'conservation of momentum'.

____

______________________________ [1 mark]

2. Maths

The diagram shows two fairground dodgem cars about to collide. Note that one of the cars is stationary before the collision occurs.

a Calculate the total momentum before the collision occurs.

Total momentum before the collision = ______ kg m/s to the right [2 marks]

b After the collision, the 80 kg car, which had been stationary, moves to the right with a velocity of 1.5 m/s.

Calculate the 80 kg car's momentum after the collision.

Momentum of 80 kg car after the collision = ______ kg m/s to the right [2 marks]

c Apply conservation of momentum to the collision and state the total momentum after the collision.

Total momentum after the collision = ______ kg m/s to the right [1 mark]

d Calculate the momentum of the 100 kg car after the collision.

Momentum of the 100 kg car after the collision ______ kg m/s to the right [1 mark]

e Calculate the velocity of the 100 kg car after the collision.

Velocity of the 100 kg car after collision = ______ m/s to the right [2 marks]

Worked Example

A 900 kg car moving at 10 m/s collides with a stationary milk float of mass 600 kg. The collision results in the milk float moving at 8 m/s in the same direction as the car was moving before the collision.

Calculate the momentum of the car after the collision.

Total momentum before the collision = 900 kg × 10 m/s = 9000 kg m/s (note: the milk float isn't moving so does not have any momentum).

So the total momentum after the collision = 9000 kg m/s (by conservation of momentum).

Momentum of the milk float after the collision = 600 kg × 8 m/s = 4800 kg m/s

So the car's momentum after the collision = 9000 − 4800 = 4200 kg m/s

3. Two rugby players are running and collide head-on with each other. The collision stops both players in their tracks and they drop to the ground. An instant after the collision their total momentum is zero so their total momentum before the collision must also be zero.

Explain how this could be possible.

______

______

______ [2 marks]

Rate of change of momentum

1. Complete the sentences about collisions. Use answers from the box.

larger	smaller

If a car's collision occurs in a shorter time interval, the rate of change of momentum is

________________ which produces a ________________ impact force. A

car's crumple zone makes the collision last for a ________________ interval which

makes the rate at which the momentum changes ________________ and results in a

________________ impact force. [5 marks]

2. **a** A 40 kg gymnast jumps down from a vaulting box onto a wooden floor and is brought to a stop in 0.2 s. Her velocity as she hits the mat is 4 m/s.

   Calculate the impact force exerted on her feet by the floor.

   Impact force = ______ N [2 marks]

   b Compare your answer to part **a** with the worked example.

   What difference does the crash mat have on the impact force on the gymnast?

   Give a reason for the difference.

   ____

   ____

   ____ [2 marks]

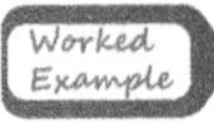

A gymnast of mass 40 kg jumps down from a vaulting box on to a gymnasium crash mat. Her velocity as she hits the mat is 4 m/s. The force exerted on her legs by the mat brings her to a stop in 0.5 s. Use the equation:

$$\text{force} = \frac{\text{change in momentum}}{\text{time taken}}$$

Maths

This equation is on the Physics Equation Sheet. You should be able to select and apply the equation – it may **not** be given in the question.

Note that Δv is the velocity change and $m\Delta v$ is the momentum change.

to calculate the impact force on her feet. The force exerted by the floor on the gymnast changes her velocity from 4 m/s to zero.

So her momentum change = $m\Delta v = 40 \times 4 = 160$ kg m/s

$$\text{Force exerted on her feet} = \frac{\text{change in momentum}}{\text{time taken}} = \frac{160}{0.5} = 320 \text{ N}$$

3. A driver is involved in a collision which brings the car to a sudden stop. The impact lasts 0.2 s. The driver's air bag inflates and cushions the driver's head so that his head comes to a stop in 1.0 s.

Use the equation

$$\text{force} = \frac{\text{momentum change}}{\text{time taken}}$$

to explain why the air bag significantly reduces the impact force on the driver's head compared with if his head had hit the steering wheel.

____

____

____ [3 marks]

4. Seat belts stop you tumbling around inside the car if there is a collision. Seat belts are also designed to stretch a bit. This means that the people inside the car take a longer time to come to a stop than the car does.

How does this affect the impact force on a person in the car?

Explain your answer.

____

____

____ [2 marks]

Transverse and longitudinal waves

1. Which **one** of the following statements is true? Tick **one** box.

Waves transfer matter but no energy. ☐

Waves transfer energy but no matter. ☐

Waves do not transfer either energy or matter. ☐ [1 mark]

2. Complete the sentences. Use words from the box.

perpendicular	parallel

In transverse waves the oscillations are ______________ to the direction of motion of the wave. In longitudinal waves the oscillations are ______________ to the direction of motion of the wave. [2 marks]

3. **a** Give **one** example of a transverse wave.

______________________________ [1 mark]

b Give **one** example of a longitudinal wave.

______________________________ [1 mark]

4. Look at the diagram of a spring below.

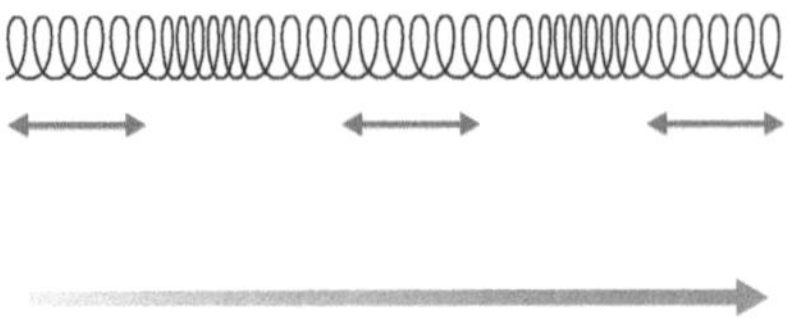

a State whether the diagram shows a longitudinal or transverse wave.

______________________________ [1 mark]

b Label on the diagram an example of a compression. [1 mark]

c Label on the diagram an example of a rarefaction. [1 mark]

5. A toy boat floats on a pond. A boy throws a stone in the water and a ripple travels across the water surface.

a State the direction in which the toy boat moves as the wave passes.

______________________________ [1 mark]

b Explain why the wave does not make the boat move closer to the edge of the pond.

____

______________________________ [2 marks]

Frequency and period

1. The diagram shows the displacement of a water wave.

On the diagram, label:

a The amplitude [1 mark]

b The wavelength [1 mark]

2. **a** Define the frequency of a wave. Give its unit.

____

Unit: ______ [2 marks]

b Define the wavelength of a wave. Give its unit.

____

Unit: ______ [2 marks]

c Define the period of a wave. Give its unit.

____

Unit: ______ [2 marks]

3. The equation below gives the relationship between the time period, T and frequency, f of a wave.

Maths

$$T = \frac{1}{f}$$

Maths
This equation is on the Physics Equation Sheet. You should be able to select and apply the equation – it may **not** be given in the question.

a Calculate the time period of a wave whose frequency is 10 Hz.

Time period = ______ s [2 marks]

b Calculate the frequency of a wave whose time period is 0.5 s.

Time period = ______ Hz [3 marks]

c Calculate the time period of a wave whose frequency is 5 kHz.

Time period = ______ s [2 marks]

Maths
Whenever you substitute a value in an equation, check that the value has standard units. If not, **convert** the value before substituting into the equation. Remember that a 'k' in front of a unit means 'kilo', so 5 kHz means 5000 Hz.

Wave speed

1. Which one of the following does wave speed depend on? Tick **one** box.

Medium (the material wave is travelling through) ☐

Amplitude ☐ [1 mark]

Maths

a Write down the formula which links wave speed, frequency and wavelength.

______________________ [1 mark]

Maths
You need to learn this equation – it may **not** be given to you in a question.

b State the units of measurement for the following quantities:

Wave speed ______

Frequency ______

Wavelength ______ [3 marks]

3. Look at the graph of a water wave below.

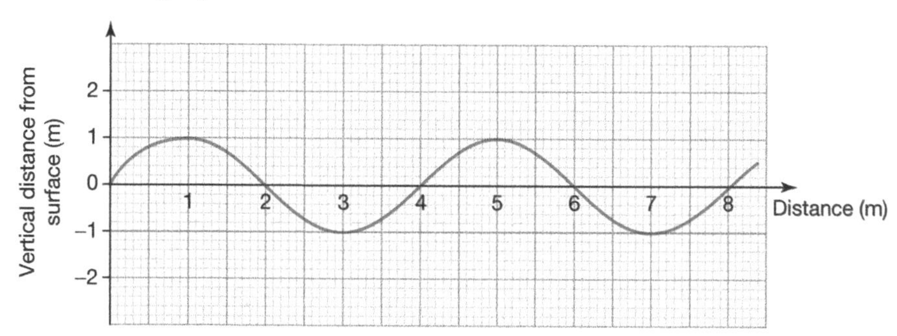

a Use the graph to determine:

The amplitude of the wave

____

The wavelength of the wave

______________________________ [2 marks]

Command word

When the question asks you to **determine**, you must use the data that is given to get your answer. Do **not** measure with a ruler unless the question tells you to measure.

b The wave is travelling at 5 m/s.

Calculate the frequency of the wave.

Use the correct equation from the Physics Equation Sheet.

Frequency = ______ Hz [3 marks]

4. 10 water waves pass a point every minute.

Maths

a Calculate the frequency of these waves.

Frequency = ______ Hz [2 marks]

b The wavelength is 2m.

Calculate the speed of the wave.

Wave speed = ______ m/s [2 marks]

5. A source produces 100 waves every 10 minutes.

Calculate the frequency of these waves in Hz. Give your answer to two significant figures.

Frequency = ______ Hz [3 marks]

6. Describe a method for determining the speed of sound in air.

Practical Include in your description:

- the measurements and calculations you would make
- why the results will not be very accurate and how you could improve the accuracy.

Practical

You should write the method in steps that follow on from each other and could be followed by someone else. You can do this by writing **first**, **next**, **then**. Or, you could use numbered points or subheadings.

____

____

____

____

____

____

____

____ [6 marks]

7. A student produces waves in a ripple tank.

Practical **a** Describe how the student could measure the frequency of the waves.

____

____ [2 marks]

b Another student is trying to measure the wavelength of the water waves with a ruler held over the waves.

Give **one** reason it is difficult to measure wavelength this way.

Suggest an improvement.

____

____ [2 marks]

c One way of working out the speed of the waves as they travel through the water is to measure the wave speed and frequency and use the equation:

wave speed = frequency × wavelength

Describe how the student could work out the speed of the waves **without** using a wavelength measurement.

Maths

Think of another equation for speed which might help here – in fact it is the equation which **defines** the quantity of speed.

____

____

____ [2 marks]

Reflection and refraction of waves

1. Practical

a Complete the diagram to show how a ray of light would be reflected at the surface of a mirror. [1 mark]

mirror

b Label the angle of reflection. [1 mark]

c Describe what apparatus you would use to produce a ray of light and measure the angle of reflection for a given angle of incidence.

____ [3 marks]

d Suggest one cause of uncertainty in the measurement of the angle of reflection.

____ [1 mark]

2. a Complete the ray diagram to show the path of the ray of light as it passes through and leaves a glass block. [3 marks]

Common misconception

You may think that the ray coming out of the block should be in a different direction to the ray entering the block. However, if the block has **parallel** sides, the change in angle as the ray enters the block is reversed when the ray leaves the block. The ray is offset, but **parallel** to the path of the original ray.

Synoptic b The glass block is replaced with a matt black object.

State what would happen to light at the surface of the block.

______________________________ [1 mark]

3. State an example of a material which causes diffuse reflection.

______________________________ [1 mark]

4. State an example of a material which causes specular reflection.

______________________________ [1 mark]

Sound waves

1. **Circle** the correct answers.

Higher Tier only

Sound waves are **longitudinal** / **transverse**.

The lower limit to the frequency range of human hearing is **2** / **20** Hz. The higher limit to the frequency range of human hearing is **2000** / **20 000** Hz. [3 marks]

2. Calculate the speed of a sound wave with a frequency of 290 Hz and a wavelength of 1.2 m.

Maths

Use the correct equation from the Physics Equation Sheet.

Speed = ______ m/s [2 marks]

3. State whether sound waves are transmitted better in gases or solids.

Synoptic

Higher Tier only

Explain your answer.

____

____ [2 marks]

Synoptic

Think about the arrangement of particles in these states of matter, and **link** this to the wave motion.

4. Describe how our ears detect sound waves in the air.

Higher Tier only

____

____ [2 marks]

5.

Synoptic

Higher Tier only

permanent magnet

electromagnet

paper cone

varying a.c. current

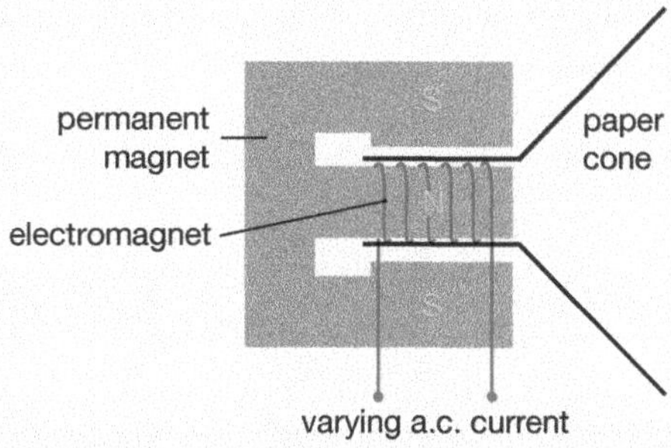

The diagram shows the parts of a loudspeaker.

Complete these sentences to explain how the loudspeaker generates sound when an a.c. signal passes through the coil.

The varying current in the coil of the electromagnet creates

____.

The magnetic field of the permanent magnet

____.

This produces a force and makes the paper cone move.

The change in the current causes

____.

The cone ____. [4 marks]

Synoptic

In the final exam some of the marks will be for **connecting** your knowledge from **different areas** of physics. To answer this question you need to **link** knowledge of the conversion of sound waves in air to vibrations of solids with ideas about electromagnetism (section 7).

6. a Describe how a loudspeaker generates sound of a higher frequency.

______________________________ [1 mark]

b Describe how a loudspeaker generates louder sounds.

______________________________ [1 mark]

Ultrasound and echo sounding

1. Complete the sentences. Use words from the box.

Higher Tier only

speeds	frequencies	absorbed	emitted
reflected	lower	higher	

Sound waves travel at different ______________ in different materials. Sound waves can either be ______________ or ______________ from boundaries between two different materials. This can allow us to detect the position of the boundary between two substances. Ultrasound is sometimes used for this. Ultrasound has a ______________ frequency than sounds we normally hear. [4 marks]

2. Ultrasound scans are used during pregnancy to check that an unborn baby is developing properly, for example, to check the size of its head.

Higher Tier only

Explain how an ultrasound emitter and receiver can be used to work out the size of different structures in an unborn baby.

____

____

____

______________________________ [4 marks]

3. Higher Tier only

State another two examples of uses of ultrasound other than for medical scans.

___

___ [2 marks]

4. Maths Higher Tier only

The speed of sound is 330 m/s in air. At the entrance of a cave a loud noise is made. Its echo is detected 2.5 s later.

Calculate the distance to the back of the cave. Show how you work out your answer.

Maths
Think very carefully about how far the sound has travelled compared to the length of the cave.

Distance = ___ [3 marks]

Seismic waves

1. Higher Tier only

Complete the sentences to compare P-waves and S-waves.

P-waves are **transverse** / **longitudinal**. P-waves travel **faster** / **slower** than S-waves.

S-waves are **transverse** / **longitudinal**. They cannot travel through **liquids** / **solids**. [4 marks]

2. Higher Tier only

a Describe the particle motion when a P-wave travels through a rock.

___ [1 mark]

b Describe the particle motion when an S-wave travels through a rock.

___ [1 mark]

3. Higher Tier only

P-waves and S-waves spread out in all directions from an earthquake.

a When seismic waves pass through the outer layer of the Earth to reach detectors a few kilometres away, P-waves arrive at the surface of the Earth before S-waves.

Give a reason why.

___ [1 mark]

b When seismic waves pass through the whole of the Earth, S-waves are not detected on the side of the Earth opposite to where the earthquake occurred. State what conclusion can be drawn from this evidence.

Explain your answer.

____

____ [2 marks]

c There is a sudden change in the direction of P-waves when they enter the Earth's core.

State what conclusion can be drawn from this evidence.

____

____ [1 mark]

The electromagnetic spectrum

1. Complete the sentences. Use words from the box.

longitudinal	transverse	frequencies
speeds	highest	lowest

Electromagnetic waves are ____________________ waves. Electromagnetic waves have a spectrum of different ____________________. Gamma rays have the ____________________ frequencies. Radio waves have the ____________________ wavelength. [4 marks]

2. Complete the sentence.

The whole of the electromagnetic spectrum has the same ____________________. [1 mark]

3. **a** Name **one** part of the electromagnetic spectrum with a smaller wavelength than ultraviolet. ____________________ [1 mark]

b Name **one** part of the electromagnetic spectrum with a lower frequency than

microwaves. ______________________________ [1 mark]

c State which has the longer wavelength: red or violet light.

______________________________ [1 mark]

4. **Maths**

The equation that links the wave speed, wavelength and frequency of a wave is

wave speed = frequency × wavelength

Use this equation to explain why parts of the electromagnetic spectrum with large wavelengths also have low frequencies.

____

______________________________ [2 marks]

Maths

You need to learn the wave equation – it may **not** be given to you in a question.

5. An electromagnetic wave has speed 3×10^8 m/s and wavelength 1×10^{-8} m.

Calculate its frequency.

Use the correct equation from the Physics Equation Sheet.

______________ [3 marks]

Reflection, refraction and wave fronts

1. **Higher Tier only**

The diagram shows the direction of travel of wave fronts of light approaching a mirror. Complete the diagram to show what will happen to the wave fronts as they leave the mirror.

mirror

[2 marks]

Remember

You can think of wave fronts as adjacent peaks of a wave – these points on the wave are moving up and down **together** at the same time and are one wavelength apart.

2. Higher Tier only

a Complete the ray diagram in the diagram to show how the wave fronts change when they enter the glass block and when they leave again.

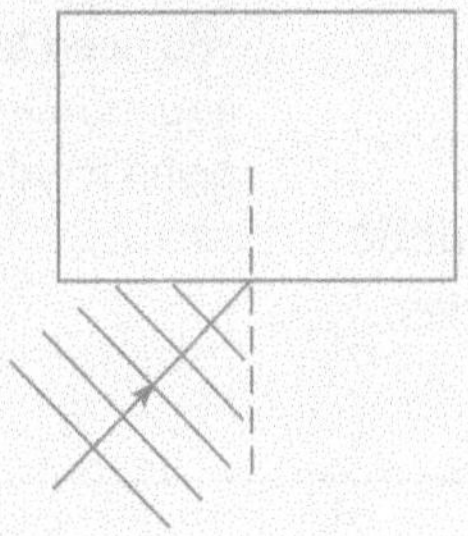

[3 marks]

b As a light ray passes into water from air, it changes direction.

State why the direction of the wave changes at the boundary between air and water.

______________________________________ [1 mark]

3. Synoptic Higher Tier only

State the effect on the following quantities when wave fronts of an electromagnetic wave enter a region where the wave speed is higher.

Wavelength ______________________

Frequency ______________________ [2 marks]

4. Maths Higher Tier only

Use the wave equation to explain your answers to question 3.

____

____

____

Common misconception

Frequency is about the number of waves produced every second – this **cannot** change after the waves have been produced.

Frequency = ______ Hz [3 marks]

Emission and absorption of infrared radiation

1. Which statement is true? Tick **one** box.

All objects absorb and emit infrared radiation. ☐

Warm objects only emit infrared radiation. ☐

Warm objects absorb and emit infrared radiation. ☐

Cool objects absorb but do not emit infrared radiation. ☐ [1 mark]

2. What does the amount of radiation emitted from a surface in a given time depend on?

Tick **one** box.

The temperature of the object ☐

The nature of the surface of the object ☐

Both the temperature and surface of the object ☐ [1 mark]

3. Practical

Hot water is placed in different coloured glass bottles of the same shape and size. One of the bottles has a matt white surface, one has a matt black surface and one has a shiny silver surface.

The temperature of the water in each bottle is recorded after 5 minutes.

a State **two** ways it can be ensured that this is a fair test.

____

______________________________ [2 marks]

b State which bottle you predict will contain water at the highest temperature after five minutes. Give a reason for your answer.

____

______________________________ [2 marks]

c State which bottle you predict will contain water at the lowest temperature after five minutes. Give a reason for your answer.

____

______________________________ [2 marks]

4. **Practical**

The diagram shows a large metal cube. One side is painted with a matt white paint, another with a matt black paint and another with a shiny black paint. The other sides are left unpainted.

The metal cube is filled with very hot water. An infrared detector is clamped at the same distance from each side. After five minutes the readings are noted.

The table shows the results.

Surface	Reading (W/m²)
bare metal	2
matt black	10
matt white	8
shiny black	8

a Which surface radiates energy most quickly per square metre? Tick **one** box.

Bare metal ☐

Matt black ☐

Matt white ☐

Shiny black ☐

[1 mark]

b Name a piece of apparatus that could be used, instead of the infrared sensor, to compare the amount of infrared radiation emitted from each surface.

______________________________ [1 mark]

c Suggest one advantage of the infrared sensor compared to the piece of apparatus you have named.

______________________________ [1 mark]

5. **Higher Tier only**

An object is absorbing more infrared radiation than it is emitting.

State what is happening to its temperature.

Give a reason for your answer.

____

______________________________ [2 marks]

Uses and hazards of the electromagnetic spectrum

1. Draw **one** line to match each part of the electromagnetic spectrum with an example of its use.

Electromagnetic wave	**Use**
Radio waves	Electric heaters, cooking food, night-vision cameras
Microwaves	Television and radio
Infrared	Energy efficient lamps, sun tanning
Visible light	Medical imaging and treatments
Ultraviolet	Satellite communications, cooking food
X-rays and gamma rays	Fibre optic communications

[6 marks]

2. State why X-rays can be used for medical imaging.

______________________________ [1 mark]

3. Name the type of electromagnetic radiation that can cause skin cancer and premature skin aging.

______________ [1 mark]

4. Name the type of electromagnetic radiation used in remote controls.

______________ [1 mark]

5. State the **three** parts of the electromagnetic spectrum that can be most harmful to humans.

State the reason why they can harm living tissue.

____

____

______________________________ [2 marks]

6. The Sun emits radiation at all wavelengths in the electromagnetic spectrum.

Higher Tier only

Only visible light, some infrared, some ultraviolet, microwaves and radio waves are transmitted through the Earth's atmosphere.

a Which property of electromagnetic waves does the transmission depend on?

______________________________ [1 mark]

b State what happens to the gamma and X-rays.

______________________________ [1 mark]

7. Radio waves and microwaves are both used to broadcast television and radio signals. Satellites are sometimes used to transmit television programmes. These television signals are sent to, and received from, the satellites using microwaves.

Higher Tier only

a State why microwaves can be used for satellite communications, but radio waves cannot.

______________________________ [1 mark]

b Firefighters use infrared cameras to help locate people in smoke-filled buildings.

Explain why it is possible to locate a person using infrared radiation, when you cannot see them using visible light.

____

______________________________ [2 marks]

8. The recommended radiation dose limit every 5 years for people who work with radiation is 0.1 Sv.

Synoptic

A spine X-ray exposes each person in the room to a dose of 1.5 mSv.

Maths

a Calculate the recommended dose limit every 5 years in mSv.

Maths
Remember that a 'm' in front of a unit means 'milli', so there are **1000** mSv in 1 Sv.

Recommended 5-year dose limit = ______ mSv [1 mark]

b State whether a patient should be worried about having an X-ray on their spine due to the radiation dose.

Give a reason for your answer.

____

____ [1 mark]

c A radiographer doing multiple spine X-rays a day is worried about the radiation dose.

Show that they are right to be concerned.

____

____ [3 marks]

d Describe how the risk to the radiographer could be reduced.

____ [1 mark]

Radio waves

1. State **two** uses of radio waves.

1 ______________________________________

2 ______________________________________ [2 marks]

2. Complete the sentences about the properties of radio waves.

Radio waves have ________________ energy and ________________ frequency compared to the rest of the electromagnetic spectrum. They have a ________________ wavelength than microwaves. [3 marks]

3. Describe how a radio transmitter produces radio waves.

Higher Tier only

____

____ [2 marks]

4. Describe how radio waves affect a radio receiver.

Higher Tier only

____

____ [2 marks]

5. State the quantity which is the same in the alternating current in a radio receiver as it is in the radio wave itself.

Higher Tier only

____

____ [1 mark]

6. Radio waves can be produced by stars.

Synoptic

State which sort of particles in the star produce the radio waves: ions or atoms.

Give a reason for your answer.

____

____ [2 marks]

Colour

1. State **two** properties which the different colours of visible light in the electromagnetic spectrum have in common.

____

____ [2 marks]

2. State two properties which are **different** for different colours of visible light.

____

____ [2 marks]

3. Explain how a blue filter turns white light into blue light.

____

____ [2 marks]

Worked Example

White light is shone on a green surface. What makes the object look green?

The green light from the white light is reflected.

All other colours are absorbed.

4. **a** Explain why a red t-shirt appears red when white light is shone on it.

____

____ [2 marks]

b State what colour a blue t-shirt would appear to be if red light was shone on it.

Explain your answer.

____

____ [2 marks]

Lenses

1. Complete the sentences. Use words from the box.

refection	refraction	thicker	thinner	real	virtual

Lenses form images using ____________________. Convex lenses bring parallel rays to a focus and are ____________________ in the middle than at the edge. Concave lenses are ____________________ in the middle than at the edge. Concave lenses always produce ____________________ images. [4 marks]

2. Explain what the 'focal length' of a convex lens is.

Include a labelled diagram in your answer.

____

____ [4 marks]

3. A lens is used to magnify a cell of width 0.3 mm. The cell appears to be 0.6 cm wide.

Maths

a State what 0.6 cm is in mm.

____

____ [1 mark]

b Calculate the magnification of this lens.

Use the correct equation from the Physics Equation Sheet.

Maths
This equation is on the Physics Equation Sheet. You should be able to select and apply the equation – it may **not** be given in the question.

Magnification = ______ [2 marks]

4. Complete the ray diagram to show how a convex lens produces the image of an object in the position shown.

Use an arrow to represent the image.

Remember
Always draw ray diagrams with a **ruler**. The rays are drawn from the tip of the object.

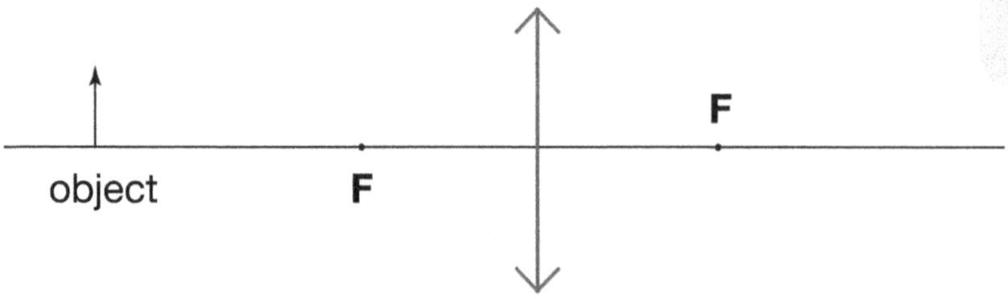

F

object F

[3 marks]

A perfect black body

1. Complete the sentences. Use words from the box.

absorbs	reflects	transmits	emitter	reflector

A perfect black body _______________ all the incident radiation on it.

That means it is also a perfect _______________. [2 marks]

2. A body is emitting more infrared radiation than it is absorbing.

State what is happening to its temperature.

Give a reason for your answer.

____

_______________ [2 marks]

3. Higher Tier only

A body is at a constant temperature.

State whether it is emitting any infrared radiation.

Give a reason for your answer.

____

_______________ [2 marks]

4. Hotter objects emit different radiation to cooler ones.

Describe **two** ways in which the radiation is different.

____

_______________ [2 marks]

Temperature of the Earth

1. Higher Tier only

Radiation from the Sun reaches the Earth's surface by passing through the Earth's atmosphere. This radiation includes short-wavelength infrared radiation.

a State two things that can happen to the radiation as it passes through the Earth's atmosphere.

1 _______________

2 _______________ [2 marks]

b State two things that can happen to the radiation when it reaches the Earth's surface.

1 ______________________________

2 ______________________________ [2 marks]

2. The Earth emits infrared radiation at a longer wavelength than the incoming solar radiation.

Explain how this causes an increase in the temperature of the Earth's atmosphere.

____

______________________________ [2 marks]

3. **Higher Tier only**

The rate of absorption of infrared radiation, the rate of emission of infrared radiation and the rate of reflection of infrared radiation to space all affect the temperature of the Earth.

a Give one example of how the temperature of the Earth can be altered by changing the radiation balance.

______________________________ [1 mark]

b For the example you have given, suggest how this factor could arise.

______________________________ [1 mark]

4. **Higher Tier only**

Explain why the Earth stays warm at night.

Draw a diagram to support your answer.

____

____

____

____

____

______________________________ [4 marks]

Magnets and magnetic forces

1. Complete the sentences. Use words from the box.

attract	repel

When two permanent magnets are close to each other, like poles of the magnets

____________ and unlike poles ____________. [2 marks]

2. Which two of the following materials are magnetic?

Tick **two** boxes.

Aluminium ☐ Nickel ☐ Cobalt ☐ Copper ☐ [2 marks]

3. A magnet is brought close to another magnet.

Describe where the force between the two magnets is largest.

________________________ [1 mark]

4. State whether magnetism is a contact or non-contact force.

Explain your answer.

____

________________________ [2 marks]

5. Describe the difference between a permanent magnet and an induced magnet.

____

____

____

________________________ [4 marks]

Analysing the question

Look at the number of marks available. It is not likely that you will be able to fit everything into a single sentence. Talk in detail about **both** types of magnet, one at a time.

Magnetic fields

1. Describe what is meant by a magnetic field.

______________________________ [1 mark]

2. Which two of the following would be attracted to the north pole of a magnet?

Tick **two** boxes.

Steel ☐

Lead ☐

The north pole of another magnet ☐

The south pole of another magnet ☐

[2 marks]

3. If you wished to pick up a steel car with a magnet, which part of the magnet would you place near to the car?

Explain your answer.

Remember
Field lines are closest together at the poles of a magnet – think about how it might apply to this question.

____

______________________________ [2 marks]

4. Magnetic field lines point in the direction of the force that would act on another north pole placed in the field at that point.

Use this information to state whether field lines point towards the **north** or **south** pole of a permanent magnet.

______________________________ [1 mark]

5. A magnetic compass points towards geographic north. However, when the compass is brought near a permanent magnet, its direction changes.

Suggest two differences between the field of the permanent magnet and the Earth's magnetic field.

____

______________________________ [2 marks]

6. Describe how to plot the magnetic field pattern of a magnet using a plotting compass.

[3 marks]

The magnetic effect of a current

1. Describe an experiment to demonstrate that a wire carrying a current generates a magnetic field. You have access to a plotting compass, a battery, and a long wire.

[3 marks]

2. **a** The side view diagram shows a wire carrying an electric current.

The shape of the magnetic field produced by the current can be shown by placing plotting compasses on the card.

Draw two magnetic field lines on the plan view diagram to show the shape of the magnetic field.

Add arrows to the lines to show the field direction.

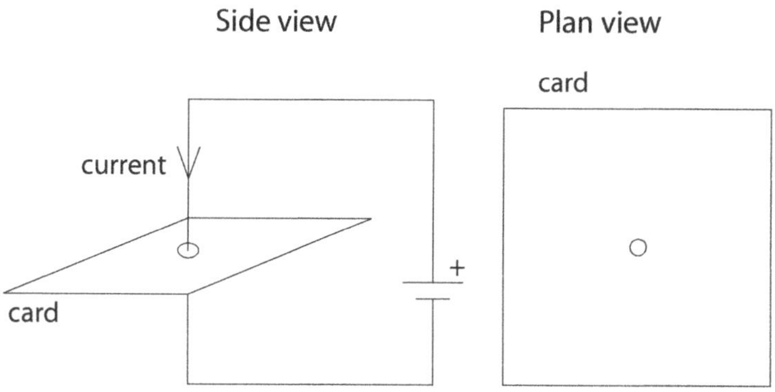

Side view

current

+

card

Plan view

card

[2 marks]

b The direction of the current is reversed.

State what happens fo the direction of the lines in the magnetic field pattern.

[1 mark]

3. A solenoid is shown in the diagram.

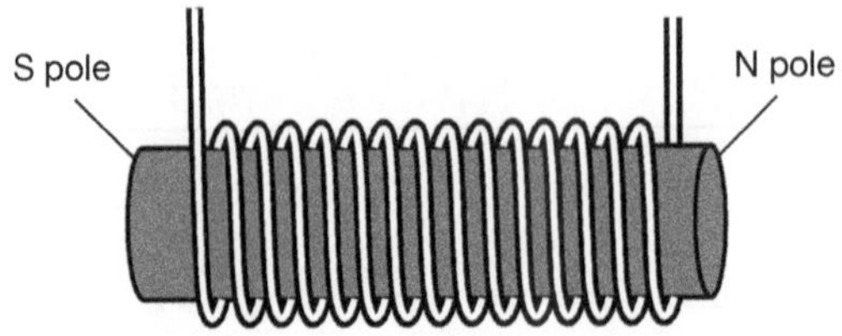

S pole

N pole

a The wire is wrapped around a material that makes the magnetic field of the soleonoid stronger.

Name this material.

____ [1 mark]

b Draw on the diagram the pattern of magnetic field lines inside and outside the solenoid. [3 marks]

4. Describe how to construct an electromagnet.

You may wish to use a diagram as part of your answer.

____

____

____ [3 marks]

5. Look at the diagram.

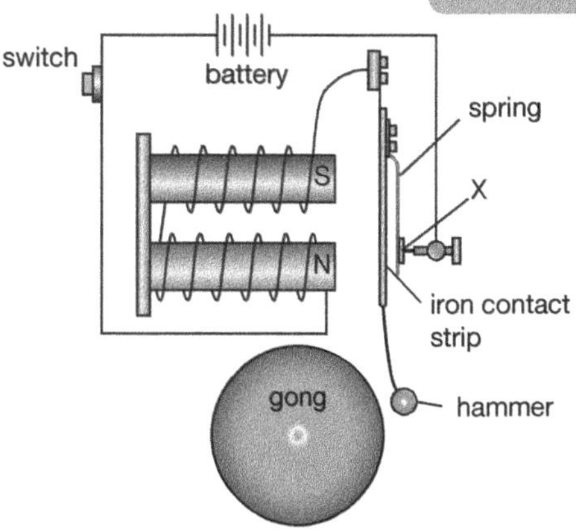

Describe what happens when the switch is pressed.

Explain the reasons for what happens.

____

____

____

____ [4 marks]

Fleming's left hand rule

1. A conductor carrying a current is placed between the poles of two magnets, at right angles to the magnetic field.

Higher Tier only

a State what happens.

____ [1 mark]

b Explain why this occurs.

____

____ [2 marks]

c State the name of the effect that this experiment demonstrates.

____ [1 mark]

2. Look at the diagram.

Higher Tier only

a On the diagram, draw an arrow to show which way the wire will move when the switch is closed.

[1 mark]

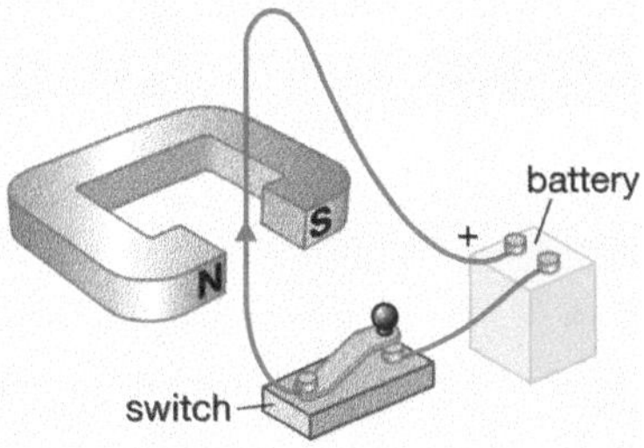

b Describe two ways in which the force on the wire could be increased.

____

______________________________ [2 marks]

Common misconception

Remember that in Fleming's left hand rule the direction of the current represented by your middle finger is conventional current, which flows from positive to negative.

Worked Example

The current in a conductor is 2 A. The conductor has length of 0.1 m and is placed in a magnetic field with flux density 0.25 T.

Calculate the force on the conductor.

Use

force = magnetic flux density × current × length

Force = 0.25 × 2 × 0.1

= 0.05 N

Maths

This equation is on the Physics Equation Sheet. You should be able to select and apply the equation – it may **not** be given in the question.

3. Higher Tier only

The current in a conductor is 3 mA and it is placed in a magnetic field with flux density 0.5 T and has a length of 2 m. The wire is at 90° to the direction of the magnetic field.

Calculate the force on the conductor.

Maths

Remember that the values you substitute in the equation must be in metres (for length) and amps (for current). 1 mA is 0.001 A.

Force = ______ N [2 marks]

4. Higher Tier only

A current-carrying conductor at right angles to a magnetic field experiences a force of 0.5 N. The conductor has a length of 10 cm and a current of 0.1 A.

Calculate the magnetic flux density of the permanent field it is in.

Force = ______ N [2 marks]

Electric motors

1. The diagram shows an electric motor. The arrows on the sides of the coil show the direction of the conventional current.

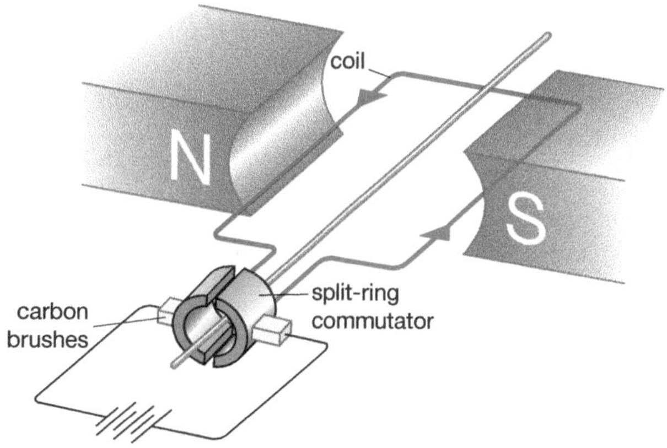

a Label the direction of the magnetic field on the diagram. [1 mark]

b Use Fleming's left hand rule to label the direction of the force on the left-hand side of the coil. [1 mark]

c Use Fleming's left hand rule to label the direction of the force on the right-hand side of the coil. [1 mark]

d State the direction of motion of the coil.

_______________ [1 mark]

e After half a turn, the coil is horizontal again.

State what needs to happen to ensure the motion continues in the same direction.

_______________ [1 mark]

f Explain how this is achieved. Refer to the diagram to help you.

____

_______________ [2 marks]

g Describe three ways in which you could make the motor spin faster.

____

____

_______________ [3 marks]

Loudspeakers

1. Headphones and loudspeakers both use magnets to work.

Higher Tier only

Which statement below is not correct? Tick one **box**.

They both use the motor effect. ☐

They both contain an electromagnet and a permanent magnet. ☐

They both use direct current. ☐

They both make molecules in the air vibrate. ☐ [1 mark]

2. The diagram shows a cross-section of a loudspeaker.

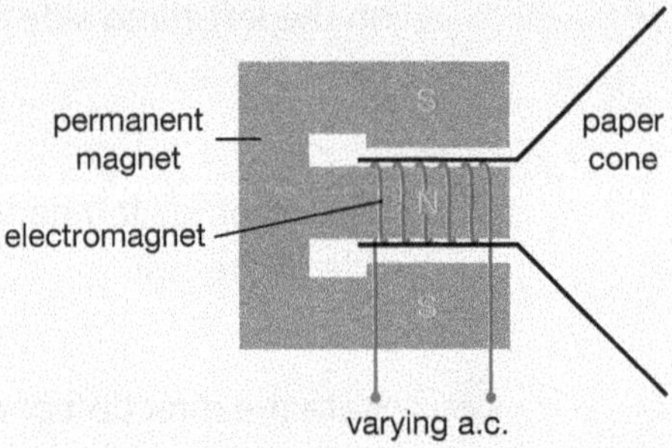

Higher Tier only

At one particular moment, the direction of the current in the wire at the top of the coil is into the page.

a State the direction of the force on the upper part of the paper cone.

____ [1 mark]

b State the direction of the current in the wire at the bottom of the coil.

____ [1 mark]

c State the direction of the force on the bottom part of the paper cone.

____ [1 mark]

d Use your answers above to explain why the field of the permanent magnet must be in opposite directions on opposite sides of the coil.

____

____ [3 marks]

e The current in the electromagnet is alternating in direction.

Describe the effect of this on the motion of the paper cone.

____

____ [2 marks]

Synoptic f Explain how this movement generates a sound.

____

____ [2 marks]

Induced potential

1. Complete the sentences. Use words from the box.

Higher Tier only

force	current	magnetic field	generator
motor	p.d.	at right angles	parallel

If a conductor is moved in a magnetic field this can induce a

______________ across the ends of the conductor. This is called

the ______________ effect. The effect only occurs if the

conductor moves ______________ to the magnetic field.

If the conductor is part of a complete circuit, an induced

______________ passes along the wire. This also

generates a ______________ which opposes the original

change that created it. [5 marks]

2. Higher Tier only

The diagram shows a wire pulled upwards through a magnetic field. The wire is connected to a meter which can measure small currents. The meter has a zero at the centre of its scale.

State how the reading on the meter would change if:

a The wire was pulled faster.

______________________________ [1 mark]

b The conductor was pushed down instead of up.

______________________________ [1 mark]

c The wire was rotated through 90 degrees then moved up.

______________________________ [1 mark]

d The magnetic field between the poles of the magnet was stronger.

______________________________ [1 mark]

3. Higher Tier only

In the diagram, a potential difference is induced when the wire is pulled upwards between the two magnets.

a State how the induced potential difference would be different if the wire was held stationary between the two magnets.

Give a reason for your answer.

____

______________________________ [2 marks]

b State how the induced potential difference would be different if the wire was stationary but the magnet was moved towards it.

Give a reason for your answer.

____

______________________________ [2 marks]

c Explain why a resistive force acts if a magnet is moved towards a stationary wire which is part of a closed circuit.

____

____ [2 marks]

Uses of the generator effect

1. Name the device which uses the generator effect to generate an a.c. output.

____ [1 mark]

2. Name the device which uses the generator effect to generate a d.c. output.

____ [1 mark]

3. **Higher Tier only** In an a.c. generator, describe how a change in the size of the force on the coil affects the output current.

____

____ [2 marks]

4. The diagram shows a dynamo. When the coil is rotated around the axle, a current is produced in the circuit and the lamp lights up.

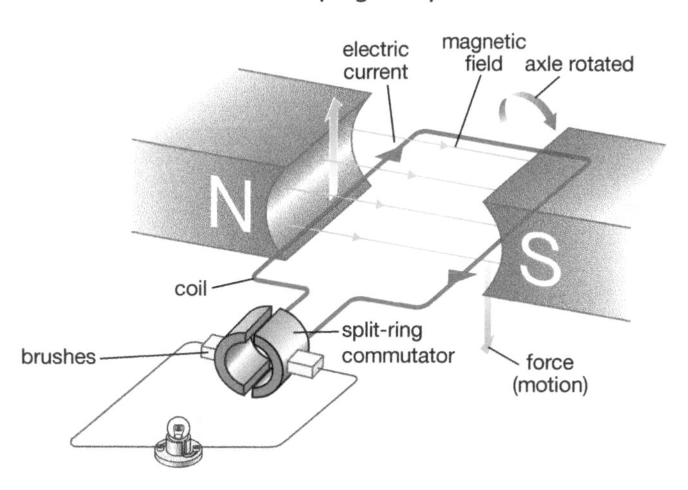

a Explain why a current is induced in the coil when the coil is turned.

____

____

____

____

____ [4 marks]

b The dynamo produces a d.c. output.

Explain how you can tell.

____

____ [1 mark]

5. On the axes, sketch the graph of the output potential difference against time as the coil is rotated smoothly. [2 marks]

Output p.d.

0

Time

Common misconception

Direct current does **not** change in direction, but the induced p.d. that causes the current **can** change in size.

Microphones

1. Higher Tier only

The diagram shows a moving-coil microphone. The energy carried by the sound waves makes the diaphragm vibrate.

sound waves

wires carrying electrical audio signal

magnet

coil

diaphragm

a Explain how the vibration of the diaphragm generates an electrical signal in the wires of the coil.

____

____ [2 marks]

b Complete the sentence.

The frequency at which the current changes direction is the frequency of

____________________. [1 mark]

2. Describe how the following factors affect the size of the p.d. induced in a coil.

a The coil has more turns.

____ [1 mark]

b The coil moves backwards and forwards with higher frequency.

____ [1 mark]

c The magnetic field is made stronger.

____ [1 mark]

3. Higher Tier only

A more sensitive microphone produces a higher induced p.d. from a sound source of given loudness.

Look at the diagram of the moving-coil microphone above.

Suggest two ways in which a moving-coil microphone could be adapted to be more sensitive to quieter sounds.

Analysing the question

Think about what needs to happen in the moving-coil microphone for it to generate a current. Then think about how this could be made more likely to happen with smaller amplitude vibrations.

____

____ [2 marks]

Transformers

1. Higher Tier only

Step-up transformers at power stations are used to transfer electrical power over large distances.

Choose one reason why. Tick **one** box.

Increasing the current through the cables reduces the energy losses in the cables. ☐

Increasing the current through the cables increases the energy losses in the cables. ☐

Decreasing the current through the cables reduces the energy losses in the cables. ☐

Decreasing the current through the cables increases the energy losses in the cables. ☐ [1 mark]

2. The diagram shows a transformer.

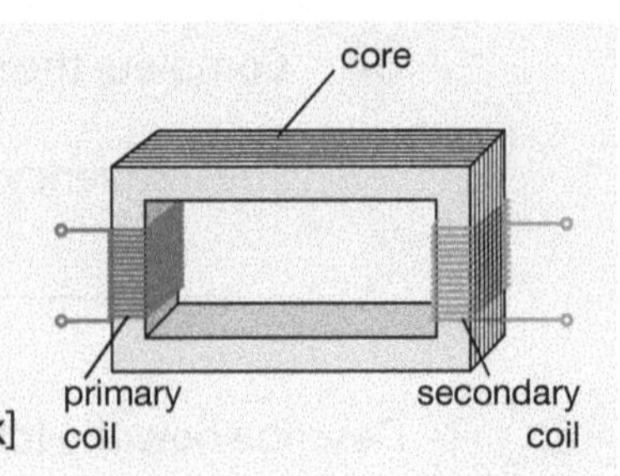

Higher Tier only

a Should an alternating or direct current be applied to the primary coil?

______________________________ [1 mark]

b Explain how the current in the primary coil induces a current in the secondary coil.

____

____

____

______________________________ [4 marks]

c Explain the purpose of the core.

______________________________ [1 mark]

d State what material the core should be made out of.

______________________________ [1 mark]

Worked Example

The primary coil p.d. in a transformer is 50 V and the secondary coil p.d. is 2 V. There are 10 turns on the secondary coil.

Calculate how many turns should be on the primary coil.

Use $\frac{V_p}{V_s} = \frac{n_p}{n_s}$

So $\frac{50}{2} = \frac{n_p}{10}$.

Multiply both sides by 10 to find n_p.

$n_p = 25 \times 10 = 250$ turns

3. **Higher Tier only**

The p.d. applied to a primary coil is 10 V and there are 5 turns on the primary coil and 50 turns on the secondary coil.

a Calculate the output p.d. across the secondary coil.

Use the correct equation from the Physics Equation Sheet.

Output p.d. = ______________________ V [3 marks]

b Is this a step-up or step-down transformer?

____ [1 mark]

c The current in the primary coil is 2 A.

Calculate the current in the secondary coil.

Use the equation

$V_s I_s = V_p I_p$

Maths

This equation is on the Physics Equation Sheet. You should be able to select and apply the equation – it may **not** be given in the question.

Current = ______________________ A [3 marks]

Our solar system

1. Complete the sentences. Use words from the box.

star	planet	galaxy

The solar system comprises of one ______________________ and eight

______________________s. Natural satellites orbit the ______________________s.

Our solar system is part of the Milky Way ______________________. [4 marks]

2. Phobos is one of the moons of Mars.

State what this means about its motion.

______________________ [1 mark]

3. Describe how the Sun was formed.

____

______________________ [2 marks]

4. Suggest what determines the mass of a star.

______________________ [1 mark]

Analysing the question
You may not have learnt this exactly, but look back at your answer to question 3 and think about what would need to be different for a more massive star.

5. **a** Describe the two forces acting on the Sun.

Include the direction of these forces.

____

______________________ [2 marks]

b Stars like the Sun are very stable.

Give one reason why.

____

______________________ [1 mark]

The life cycle of a star

1. State the process by which stars release energy.

______________________ [1 mark]

2. Describe what causes a star to end its main-sequence phase.

______________________ [1 mark]

3. Explain how all the elements were formed.

____

____

______________________ [3 marks]

4. Could very heavy elements have formed from stars similar to our Sun? Give a reason for your answer.

____

______________________ [2 marks]

5. Describe three similarities in the life cycles of low-mass and high-mass stars.

____

____

______________________ [3 marks]

6. Describe three differences between the end of the star's life for low-mass and high-mass stars.

____

____

____

______________________ [3 marks]

Command words

In a **describe** question you need to recall and use **scientific** language. Read the question carefully and make a list of key words that come to mind.

Orbital motion, natural and artificial satellites

1. Name the force which keeps the Earth in orbit around the Sun.

____________________ [1 mark]

2. Describe the difference between natural and artificial satellites.

Give an example of each.

____

____

____________________ [3 marks]

3. Describe the difference between a planet and a moon.

____________________ [1 mark]

4. **Higher Tier only**

Explain how the Earth can have a steady speed in its orbit round the Sun but a constantly changing velocity.

____

____________________ [2 marks]

5. **Higher Tier only**

The table shows the orbital speeds of four satellites in orbit round the Earth.

Satellite	Orbital speed (km/s)
International Space Station	7.66
communication satellite	3.07
Hubble space telescope	7.50
GPS satellite	3.89

a Complete the sentence.

To stay in its orbit around the Earth, each satellite must move at a particular

____________________. [1 mark]

b State what would happen to the orbit of each satellite if the orbital speed changed.

Give a reason for your answer.

____

____ [2 marks]

Red-shift

1. Most galaxies are moving away from the Earth.

a Describe how we can tell this from the light they emit.

____ [1 mark]

b What is the name of this effect?

__________________ [1 mark]

More distant galaxies are moving away more quickly.

c Describe how we can tell this.

____ [1 mark]

d State what this suggests about the position of these galaxies in the past.

____ [1 mark]

2. **a** Describe the Big Bang theory.

____

____ [2 marks]

b Explain how red-shift from distant galaxies provides evidence for the Big Bang theory.

___

___

___ [3 marks]

3. **a** Very distant galaxies have very large red-shifts.

What objects in very distant galaxies have provided evidence that these distant galaxies are further away than expected? Tick **one** box.

Black dwarf ☐ Black holes ☐

Red super giants ☐ Supernovae ☐ [1 mark]

b What can astronomers conclude from these objects about the expansion of the Universe?

___ [1 mark]

Dark matter and dark energy

1. Red-shift tells us how fast a galaxy is receding from the Earth. Astronomers have discovered that supernovae in very distant galaxies have slightly less red-shift than expected. This is because they are further away than their measured red-shift suggests.

a What can astronomers conclude about the lower-than expected red-shift of these distant objects?

Tick **one** box.

A mysterious force has pushed galaxies together, making the expansion slow down. ☐

A mysterious force has pushed galaxies apart, making the expansion slow down. ☐

A mysterious force has pushed galaxies together, making the expansion speed up. ☐

A mysterious force has pushed galaxies apart, making the expansion speed up. ☐

Scientists do not understand the reason for this effect. ☐

[1 mark]

b State the name they have given to this mysterious mechanism.

______________________________ [1 mark]

2. Scientists also cannot explain the mass of galaxies based on the amount of matter they know exists in stars. They have worked out there must be more mass than is visible in the stars.

b State the name given to this invisible mass.

______________________________ [1 mark]

b The invisible mass in galaxies has not been directly observed. However, it can be detected indirectly.

Complete the sentence to describe how this invisible mass can be detected.

Visible objects are acted on by a force due to the ______________________________ force caused by the object they cannot see. [1 mark]

3. Matter that can be seen, such as stars, planets, gas and dust, makes up only 4% of the mass of the Universe. It is estimated that this mass is about 10^{53} kg.

Estimate the mass of the entire Universe.

Give your answer to the nearest order of magnitude.

Maths

To multiply two numbers in standard form multiply the whole numbers first. Next, multiply together the two powers of ten. Remember, when you multiply powers of ten you add the indices.

Mass of Universe = ______________________ kg [3 marks]

Answers

Section 1: Energy changes in a system

Energy stores and systems

1. Walking along a flat road [1 mark]
2. [1 mark for each correct line. If more than one line is drawn from a system, no marks for that system.]

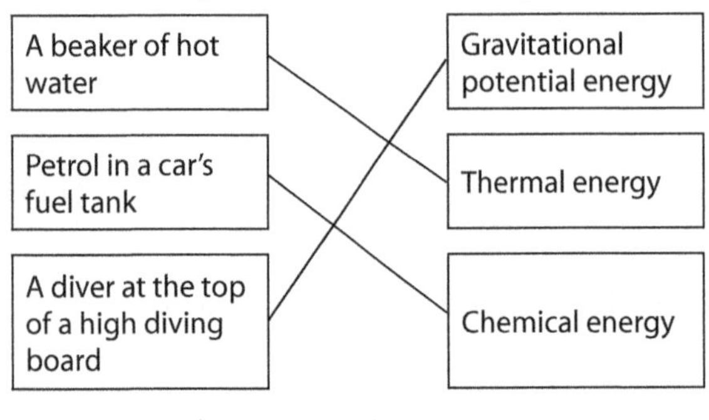

3.

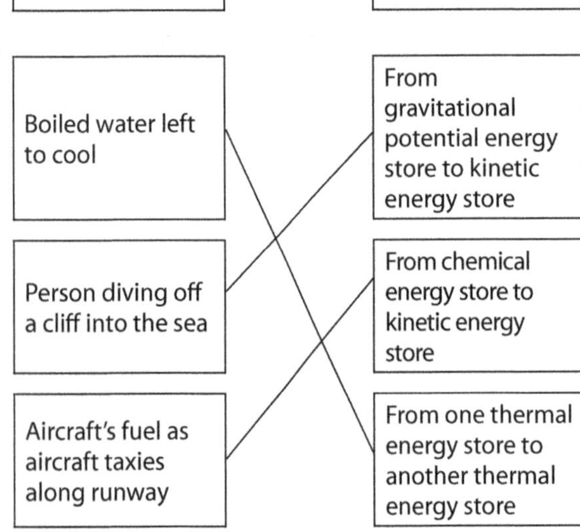

4. Elastic potential, kinetic, kinetic, thermal [1 mark each]

Calculating energy changes

1. a) $E_k = \frac{1}{2}mv^2$ or kinetic energy = 0.5 × mass × speed²
   b) $\frac{1}{2} \times 900 \times (4)^2$ [1 mark] = 7200 J [1 mark]
2. a) GPE = mgh [1 mark]
   b) 2.0 × 9.8 ×2.0 [1 mark] = 39.2 (J) [1 mark]
3. a) 50 × 9.8 × 5.0 [1 mark] = 2450 [1 mark] unit: J [1 mark]
   b) 2350 J [1 mark]
4. a) $\frac{1}{2} \times 800 \times 10^2$ [1 mark] 40 000 (J) [1 mark]
   b) $\frac{1}{2} \times 800 \times 5^2$ [1 mark] 10 000 (J) [1 mark]
   c) 40 000 J – 10 000 J = 30 000 J [1 mark]
   d) Energy transferred as thermal energy [1 mark] to the brakes / surroundings. [1 mark]
5. $\frac{1}{2} \times 20 \times (0.1)^2$ [1 mark] 0.1 (J) [1 mark]
6. a) 0.12 m (or 12 cm) [1 mark]
   b) $\frac{1}{2} \times 25 \times 0.12^2$ [1 mark] 0.18 J [1 mark]
7. a) At point of release: elastic potential energy to kinetic energy. [1 mark] Stone travelling upwards towards its maximum height: kinetic energy to gravitational potential energy. [1 mark] Stone falling back towards ground: gravitational potential energy to kinetic energy. [1 mark]
   b) 1.0 = 0.05 × 9.8 × h [1 mark] so height = 1.0/(0.05 × 9.8) [1 mark] = 2.04(08) [1 mark] = 2.0 (to 2 sig figs) [1 mark]

Calculating energy changes when a system is heated

1. 0.5 × 4200 × 80 [1 mark] 168 000 J (or 168 kJ) [1 mark]
2. a) 23 °C [1 mark]
   b) $c = \frac{\Delta E}{m\Delta\theta}$ [1 mark] $= \frac{21260}{1 \times 23}$ [1 mark] 924 [1 mark]
   c) Reduce store of thermal energy transferred to surroundings [1 mark]
   d) Some thermal energy store transferred to surroundings / used to raise temperature of immersion heater [1 mark]
3. a) Gravitational potential, kinetic, gravitational potential [1 mark each]
   b) 250 × 9.8 × 60 [1 mark] = 147 000 J [1 mark]
   c) 250 × 790 × 0.5 [1 mark] = 98 750 J [1 mark] 99 000 J [1 mark]
4. a) 1.0 × 4200 × 80 [1 mark] 336 000 J [1 mark]
   b) 0.5 × 500 × 80 [1 mark] = 20 000 J [1 mark]
   c) Total energy supplied = 336 000 + 20 000 = 356 000 J [1 mark]

Work and power

1.

The force of gravity pulling a free-wheeling cyclist faster and faster down a hill

A person pushing a heavy shopping trolley to get it moving

A person lifting a box from the floor onto a shelf

From chemical energy store to kinetic energy store

From chemical energy store to gravitational potential energy store

From gravitational potential energy store to kinetic energy store

2. a) 10 × 15 [1 mark] = 150 J [1 mark]
   b) 150/10 [1 mark] = 15 W [1 mark]
3. a) 4000 × 50 [1 mark] = 200 000 J [1 mark]
   b) 200/5 [1 mark] 40 kW [1 mark] (or 200 000/5 and kW conversion)
4. a) $mgh = 800 \times 9.8 \times 10$ [1 mark] = 78 400 J [1 mark]
   b) $\frac{78\,400}{20}$ [1 mark] = 3920 W [1 mark]

Conservation of energy

1. Dissipation [1 mark]
2. Energy can be destroyed [1 mark]
3. 165 J [1 mark]
4. Decreases (to zero), force of friction (between ground and rotating wheels), same / constant / conserved, thermal energy store of wheels and surroundings [1 mark each]

Ways of reducing unwanted energy transfers

1. Lubricate wheels [1 mark]
2. a) Marks in three bands according to level of response.

Level 3 [5–6 marks]: Method described in detail and steps in an order that makes sense. Dependent and control variables correctly identified. Method would lead to the production of valid results.
Level 2 [3–4 marks]: Method described with mostly relevant detail although steps may not be in logical order. Some detail may be missing.
Level 1 [1–2 marks]: Simple statements made about some of relevant parts of method but steps may not be in logical order and would not lead to production of valid results.
Level 0: No relevant content.
Points that should be made: • Place small beaker inside large beaker. • Pack insulating material in gap between beakers. • Pour hot water into small beaker. • Control volume of hot water (using measuring cylinder). • Cover with lid and slot thermometer through hole in lid into water. • Control initial temperature (always start timing at same initial temperature). • Measure water temperature at fixed time intervals / every 2 minutes. • Repeat with different materials, using a constant thickness of insulator (i.e. fill gap between the two beakers each time).

   b) Expanded polystyrene, [1 mark] line for expanded polystyrene above line for glass fibre once cooling starts [1 mark]

Efficiency

1. a) Chemical, kinetic, thermal [1 mark each]
   b) 580 J [1 mark]
   c) 0.42 [1 mark] 42% [1 mark]
2. 0.8 = useful output energy/500 MJ [1 mark] useful output energy = 0.8 × 500 MJ [1 mark] = 400 MJ [1 mark]
3. 0.5 = 1000 MJ/useful input energy transfer [1 mark] useful input energy transfer = 1000 MJ/0.5 [1 mark] = 2000 MJ [1 mark]
4. Any two of: loft insulation, cavity wall insulation, preventing draughts, double glazing [1 mark each]

National and global energy resources

1. a) Coal and gas [1 for both]
   b) Uranium [1 mark]
   c) Energy resource is finite / will eventually run out [1 mark] because not replenished [1 mark]
   d) Any three from [1 mark each]:
      - Biggest changes between 2013 and 2015 involve coal and renewables.
      - Percentage of electricity generated by coal decreased by 15%.
      - Percentage of electricity generated by renewables increased by over 10%.
      - Only small changes / no trend for gas and nuclear.
   e) Any three from [1 mark each]:
      - Burning coal causes emission of carbon dioxide / contributes to climate change.
      - Use of nuclear fuel does **not** result in carbon dioxide emission / does **not** cause climate change.
      - Burning coal produces soot particles / pollution / acid rain from sulfur dioxide.
      - Use of nuclear fuel produces radioactive waste.
      - Radioactive waste will be hazard for hundreds of years.
2. a) Any one of: wood chips, wood waste, agricultural waste, waste from sugar cane processing [1 mark]
   b) So electricity always available / to avoid power cuts [1 mark]
   c) Both unreliable sources. Wind: no wind = no electricity, [1 mark] solar: no electricity when dark [1 mark]
3. Marks in three bands according to level of response.

Level 3 [5–6 marks]: Clear coherent description of trends in energy consumption over time for industry, transport and homes involving several comparisons. Logical possible explanation for why trends occurred.
Level 2 [3–4 marks]: Some valid comparisons made of trends in energy consumption over time. Some valid suggestions made, with reasons. There may be some incorrect or irrelevant points.

Level 1 [1–2 marks]: Major trends in energy consumption for industry, transport and homes identified but no direct comparisons. Suggested explanations are vague with few correct and relevant points.

Level 0: No relevant content.

Points that should be made:

- Largest increase between 1970 and 2014 is in energy consumed by transport.
- Increase in population means more vehicles on the road, more vehicles from Europe since joining EU, more people driving.
- Largest decrease between 1970 and 2014 is in energy consumed by industry.
- Industry more efficient in its use of energy, less heavy industry in UK in 2014 than in 1970.
- Small increase in the energy consumed by homes between 1970 and 2014.
- Increasing population (and household income) means more energy consumed.
- In recent years there has been a reduction in consumption by transport and homes.
- Appliances used in homes and vehicles are more efficient.

4. Any 2 from [1 mark each]:
- UK heating systems / appliances / vehicles / industry now more efficient.
- Many other countries becoming more industrialised.
- Industry in developing countries may not use energy efficiently.
- Large increase in number of cars / vehicles used around the world.

Section 2: Electricity

Circuit diagrams

1.

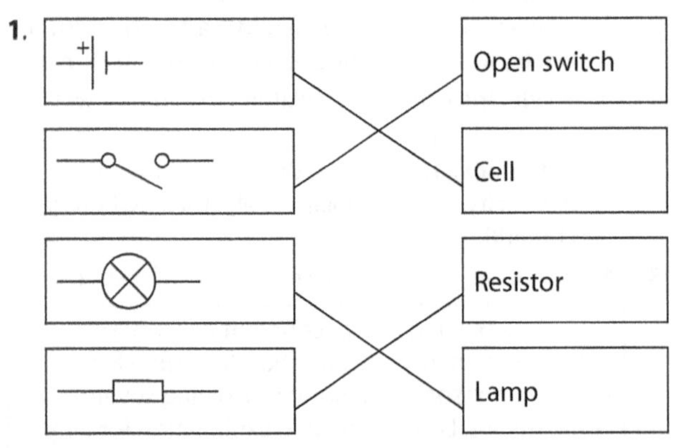

2. Four [1 mark]

3. a) X [1 mark] **b)** Z [1 mark]
c) Y [1 mark] **d)** W [1 mark]

Electrical charge and current

1. Same, [1 mark] current same at any point in a circuit made up of a single (closed) loop. [1 mark]

2. a) Charge [1 mark]
b) Cell [1 mark]
c) [2 marks for voltmeter connected across lamp]

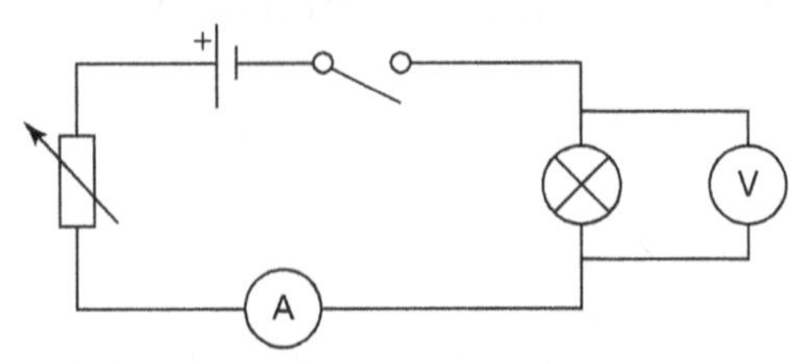

d) 0.15×20 [1 mark] = 3.0, [1 mark] coulomb or C [1 mark]

e) $\frac{1}{20 \times 10^{-3}}$ [1 mark] 50 (s) [1 mark]

Electrical resistance

1.

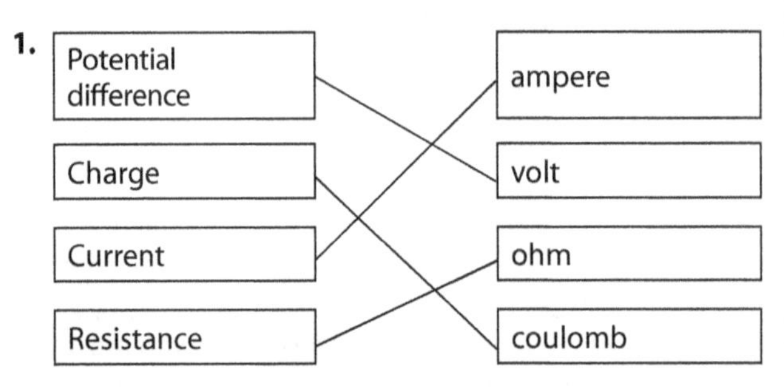

2. [1 mark each for cell, ammeter and voltmeter correctly positioned]

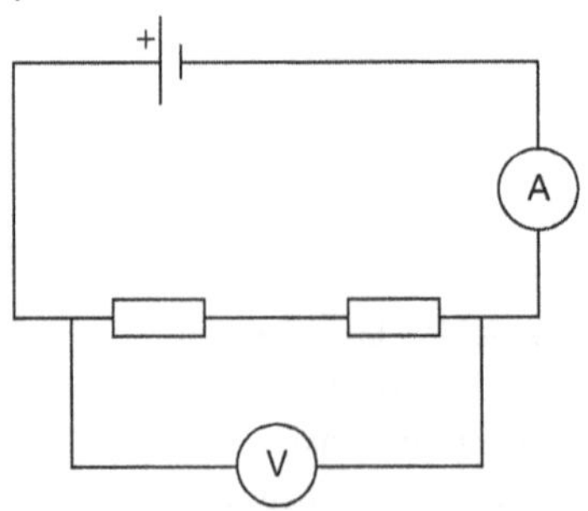

3. a) 0.25×6 [1 mark] 1.5 V, [1 mark]
b) Ammeter reading / current would decrease [1 mark]
c) $\frac{1.5}{48 \times 10^{-3}}$ [1 mark] = 31.25 Ω [1 mark] 31 Ω [1 mark]

4. a) 4.47, 5.60, 6.71, 7.84, 8.98, 9.93 [2 marks for 6 answers correct] 3 significant figures [1 mark]
b) Same thickness / diameter of wire / same temperature [1 mark]
c) Spot anomalies, [1 mark] reduce effect of random errors [1 mark]

d) x-axis: length (of wire), [1 mark] y-axis: resistance [1 mark]

e) Open switch between measurements [1 mark]

Resistors and I–V characteristics

1. a) Variable resistor [1 mark]

b) [1 mark for variable resistor symbol connected between X and Y]

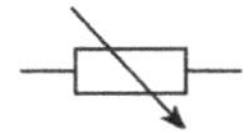

c) Reverse battery connections [1 mark]

d) Current is directly proportional to p.d., [1 mark] graph = straight line, [1 mark] passing through origin [1 mark]

2. a) Resistance, [1 mark] temperature [1 mark]

b) Non-linear [1 mark]

c) Large [1 mark]

3. Marks in three bands according to level of response.

Level 3 [5–6 marks]: Detailed and coherent plan with all major steps described in correct logical sequence. Plan addresses issues of obtaining repeatable measurements and negative values of current and p.d.
Level 2 [3–4 marks]: Detailed and coherent plan with all major steps described in correct logical sequence. Some detail may be missing.
Level 1 [1–2 marks]: Simple steps related to relevant apparatus, steps may not be given in correct logical sequence.
Level 0: No relevant content.
Points that should be made: • Current through X measured using ammeter. • p.d. across X measured using voltmeter. • Repeatedly adjust variable resistor to change current through X or change p.d. (across X). • Connections to the battery are reversed. • Obtain a range of negative current and p.d. readings. • Repeat procedure to obtain repeat readings of current for each p.d. value. • Average repeat current values. • Plot line graph of current against p.d. • If line of best fit can be drawn that closely matches all data points, this indicates a proportional relationship.

4. a) p.d. [1 mark]

b) $\frac{1.51}{3.52 \times 10^{-3}}$ [1 mark] = 492 [1 mark]

c) $\frac{1.51}{1.23 \times 10^{-3}}$ [1 mark] = 1230 [1 mark]

d) Resistance lower [1 mark] as LDR receives increasing amounts of light [1 mark] (accept converse description) [allow resistance changes depending on the amount of light that hits the LDR for 1 mark]

5. a) Thermistor symbol shown connected between X and Y [1 mark]

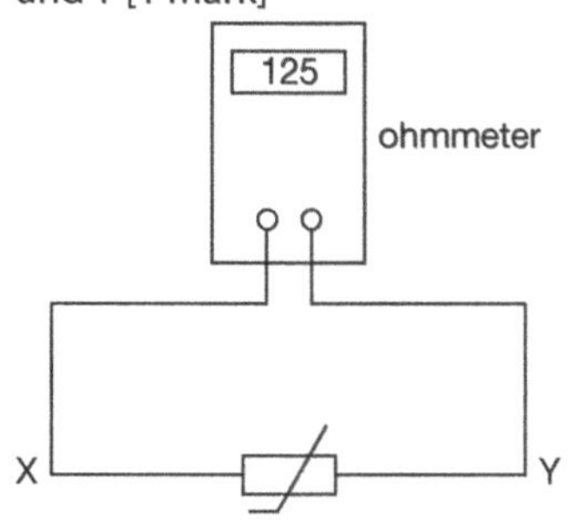

b) Resistance decreases as thermistor's temperature increases [2 marks] [Allow only 1 mark for temperature of the thermistor changes as its resistance changes.]

Series and parallel circuits

1. a) Resistors in series, [1 mark] same current through all three resistors. [1 mark]

b) 6.6 Ω [1 mark]

c) 0.5 V [1 mark]

2. a) All three resistors have same p.d. across them, [1 mark] resistors in parallel [1 mark]

b) $\frac{1.5}{5}$ [1 mark] = 0.3 A [1 mark]

3. a) $\frac{1.5}{15}$ [1 mark] = 0.1 A [1 mark]

b) $\frac{1.5}{10}$ [1 mark] = 0.15 A [1 mark]

Mains electricity

1. a) 50 Hz [1 mark]

b) 230 V [1 mark]

2.

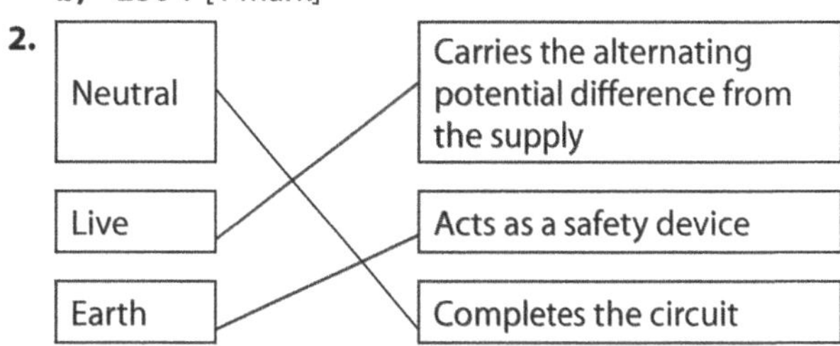

3. Alternating, brown, earth [1 mark each]

4. Any 3 of: if the fault occurs, there is a p.d. between metal casing and ground / earth, [1 mark] if person touches casing, current could pass through person, [1 mark] earth wire carries current to ground instead, [1 mark] because of very low resistance, [1 mark] protecting person from electric shock / preventing current passing through person. [1 mark]

Energy changes in circuits

1. Kinetic, thermal [1 mark each]

2. Operate vacuum cleaner for longer period of time, [1 mark] use vacuum cleaner with higher power rating [1 mark]

3. 2000 × 30 × 60 [1 mark] = 3 600 000 J [1 mark]
4. a) 1000 × 5 × 60 [1 mark] = 300 000 J [1 mark]
   b) $\frac{300000}{230}$ [1 mark] = 1304 C [1 mark]
5. a) 2 × 10 × 60 [1 mark] = 1200 J [1 mark]
   b) $\frac{1200}{3}$ [1 mark] = 400 C [1 mark]

Electrical power

1. 230 × 2.0 [1 mark] = 460 [1 mark] W [1 mark]
2. $P = I^2R$ [1 mark] = $4.0^2 \times 0.4$ [1 mark] = 6.4 W [1 mark]
3. Microwave A lower power (600 W), [1 mark] microwave B higher power (900 W), [1 mark] so microwave B transfers energy more quickly / at higher rate. [1 mark]
4. a) 230 × 0.40 [1 mark] = 92 W [1 mark]
   b) 92 × 10 × 60 [1 mark] = 55 200 J [1 mark]

The National Grid

1. When all power stations are connected, the National Grid can supply electricity to any town / city. If each town / city depended on its own power station then if there was a problem with the power station the town / city would have a power cut. [1 mark]
2. Increases, decreases, [2 marks] decreases, increases [2 marks]
3. Transformer A is a step-up and transformer B is a step-down [1 mark]
4. The National Grid transmits electricity at high voltages to improve efficiency. [1 mark] The National Grid transmits electricity at low currents to improve efficiency. [1 mark]

Static electricity

1. Sphere A is positive, sphere B is positive. [1 mark] Sphere A is negative, sphere B is negative. [1 mark]
2. a) Electrons transferred from cloth to rod [1 mark]
   b) Cloth becomes positively charged, [1 mark] because it has fewer electrons / loses electrons [1 mark]
   c) Rods move towards each other [1 mark]
   d) Attractive force between rods [1 mark] because rods are carrying opposite charges. [1 mark]

Electric fields

1. a) [1 mark]

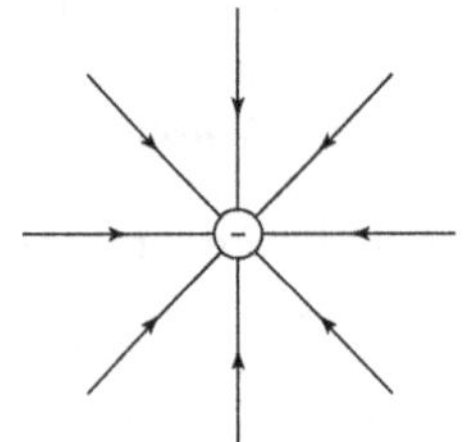

   b) Force gets bigger [1 mark]
2. Towards sphere [1 mark]
3. a) 75 MJ [1 mark]
   b) 30 000 A [1 mark]
   c) 170 Ω [1 mark]

Section 3: Particle model of matter

Density

1. Density = $\frac{\text{mass}}{\text{volume}}$ or $\rho = \frac{m}{V}$ [1 mark]
2. 740 = 3.7/mass [1 mark] mass = 3.7/740 [1 mark] 0.005 [1 mark] m^3 [1 mark]
3. a) Micrometer [1 mark]
   b) Measuring cylinder [1 mark]
4. a) Marks in three bands according to level of response.

Level 3 [5–6 marks]: Method described in detail including equation required to calculate density. Steps in an order that makes sense and could be followed by someone else to obtain valid results. Some valid suggestions about reducing errors are made, with reasons.
Level 2 [3–4 marks]: Method described clearly but some detail may be missing and steps may not be in an order that makes sense. At least one valid suggestion about reducing errors made, with reasons.
Level 1 [1–2 marks]: Basic description of measurements needed with no indication of how to use them or reduce errors.
Level 0: No relevant content.
Points that should be made: • Measure mass using a balance. • Calculate density using $\rho = m/V$. • Fill displacement can with water until full up to spout. • Submerge rock in water. • Measure volume / mass of water displaced (with a measuring cylinder). • Volume of water displaced = volume of rock. • Do repeats and calculate mean. • Reference to avoiding systematic errors (e.g. reading measuring cylinder at eye level or taking account of meniscus or ensuring balance is zeroed).

   b) Marks in two bands according to the level of response.

Level 2 [3–4 marks]: Evidence set out clearly to reach a conclusion. Correct density calculated for sample and correct conclusion drawn.
Level 1 [1–2 marks]: Correct use of equation and attempt to compare to one of the given densities. Response may not be structured coherently.
Level 0: No relevant content.

Points that should be made:

- Calculate density using $\rho = m/V$
- Correct value 3600 kg/m³ (2 sig figs).
- Data show that sample is more likely to be a meteorite as it is closer to this density range.
- Outside expected density range for terrestrial rock.

5. Marks in two bands according to level of response.

Level 2 [3–4 marks]: Method described in detail including equation required to calculate density. Steps in an order that makes sense and could be followed by someone else to obtain valid results.
Level 1 [1–2 marks]: Basic description of measurements needed with no indication of how to use them.
Level 0: No relevant content.

Points that should be made:

- Need to work out mass and volume as density = mass/volume.
- Put an empty measuring cylinder on some scales and note its mass.
- Add some of liquid and note new mass.
- Take away new mass from mass of empty measuring cylinder to work out mass of liquid.
- Divide mass of liquid by volume in measuring cylinder.

a) Very few circles / do not occupy more than about one third of box, [1 mark] no circles touching [1 mark]

b) A gas bubble displaces a volume of liquid equal to its own volume, [1 mark] weight of gas bubble much less than weight of same volume of water (because less dense), [1 mark] causing resultant force (upthrust) on gas bubble. [1 mark]

Changes of state

1. a)

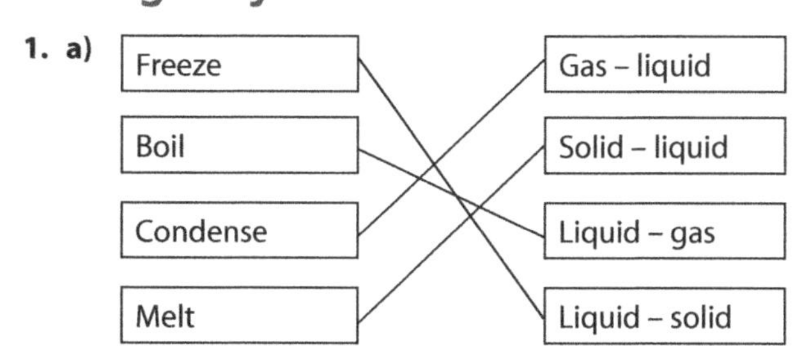

b) Boil, [1 mark] melt [1 mark]

2. Changes state from solid to gas with no liquid phase. [1 mark]

3. The change is reversible / material has its original properties if change is reversed. [1 mark]

4. Burning [1 mark]

5. Student B, [1 mark] because no particles lost / gained [1 mark]

Specific heat capacity and latent heat

1. In hot gas: faster / higher speed / more k.e. [1 mark]

2. Kinetic energy [1 mark]

3. a) Particles moving more quickly on average. [1 mark]

b) Bath has far more particles / greater mass [1 mark] so total energy of particles greater. [1 mark]

4. Despite having same temperature so same kinetic energy, [1 mark] water particles have greater internal energy because energy was supplied / gained for change of state to occur. [1 mark]

5. Energy is used to change state of material [1 mark] instead of increasing kinetic energy store of particles. [1 mark]

6. a)

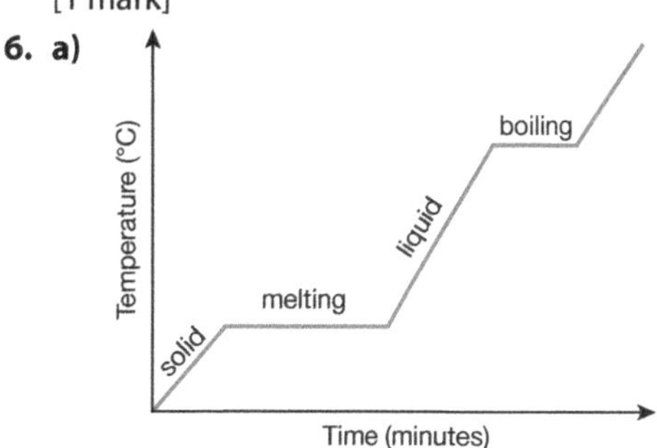

b) Melting

c) $5 \times 45\,000$ [1 mark] $= 225\,000$ [1 mark]

7. $15\,000 = 0.5 \times L$ [1mark] $L = 15\,000/0.5$ [1 mark] = 30 000 J/kg [1 mark]

Particle motion in gases

1. Due to force of collisions with walls of container. [1 mark]

2. They move more / faster / with more kinetic energy [1 mark]

3. Pressure will increase [1 mark] because more frequent collisions, [1 mark] more force in collisions [1 mark]

4. a) Bung will pop out or flask will shatter [1 mark]

b) As temperature increases, particles move around more quickly [1 mark] resulting in increased pressure. [1 mark]

5. Marks in two bands according to the level of response.

Level 2 [3–4 marks]: Detailed explanation with logical links between clearly identified relevant points.
Level 1 [1–2 marks]: Relevant, but separate, points are made. Logic may be unclear.
Level 0: No relevant content.

Points that should be made:

- Hot jam heats the layer of air above.
- Layer of hot air trapped by lid / no air can escape.
- Initially high gas pressure.
- Heated air cools over time.
- This lowers gas pressure (same volume).
- This means there is greater pressure on top of lid (from atmospheric pressure) than underneath.
- This makes jar difficult to open / so more force needed.

Increasing the pressure of a gas

1. Marks in two bands according to level of response.

Level 2 [3–4 marks]: Detailed explanation with logical links between clearly identified relevant points.
Level 1 [1–2 marks]: Relevant, but separate, points are made. Logic may be unclear.
Level 0: No relevant content.

Points that should be made:

- Larger volume means same number of particles in larger space.
- Air particles in constant motion.
- Air pressure caused by force of collision of air particles with wall of container.
- Fewer collisions of air particles with container if container is bigger.
- So lower pressure.
- Lower pressure inside means less risk of container bursting.

2. pV before = 100 000 × 0.001 [1 mark] = 100, [1 mark] pV after = 100, [1 mark] final pressure = 100/0.0002 [1 mark] = 500 000 [1 mark]
3. No, because energy is being added to system / work done, [1 mark] this increases internal energy of the gas. [1 mark]

Section 4: Atomic structure

Protons, neutrons and electrons

1. [1 mark for each correct entry]

proton	positive	**1**	**inside**
neutron	**neutral**	**1**	**inside**
electron	**negative**	negligible	outside nucleus

2. Nucleus, further, lower [1 mark each]
3. Absorbed some radiation [1 mark] so electrons are further from nucleus / occupy a higher energy level [1 mark]

The size of atoms

1. 10^{-5},10^{1},10^{3},10^{4} first and last correct, [1–2 marks] all correct [all 4 marks]
2. 2×10^{-3} m [1 mark]
3. A football pitch [1 mark]
4. **a)** A [1 mark] **b)** C [1 mark]

Elements and isotopes

1. Neutral, electrons, protons, atomic, protons / neutrons, neutrons / protons [1 mark each]

Atomic number	Mass number	Number of protons	Number of neutrons
6	12	6	**6**
11	**13**	**11**	12
27	59	**27**	32
13	**27**	13	14

3. Atoms of same element, [1 mark] with same number of protons / same atomic number [1 mark] but different numbers of neutrons / different mass number [1 mark]
4. **a)** 27, 33, 27 [1 mark each]
   b) $^{60}_{27}$Co [1 mark]
   c) $^{59}_{27}$Co [1 mark]

Electrons and ions

1.

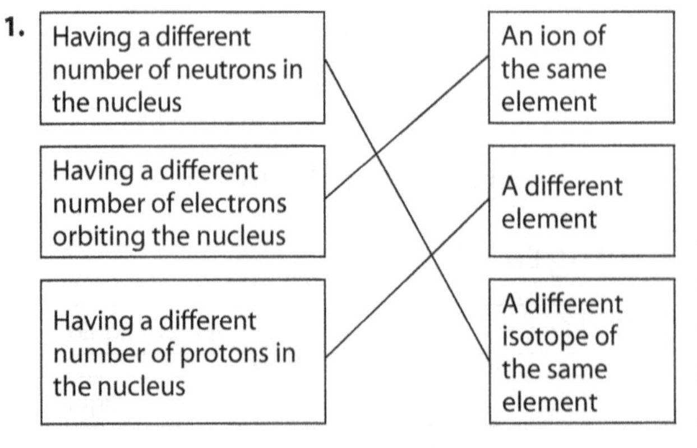

2. Positively charged [1 mark] because atoms neutral / equal numbers of positive and negative charge [1 mark] so losing electron (negatively charged) means more protons than electrons. [1 mark]
3. 17, 18, 18 [1 mark each]
4. 13, 14, 11 [1 mark each]

Discovering the structure of the atom

1. Discovery of electron [1 mark]
2. **a)** Ball / sphere of positive charge [1 mark] with negative electrons embedded throughout [1 mark]

b) Concentrated in a small space [1 mark]

3. Instead of electrons surrounding the nucleus anywhere in a cloud, [1 mark] Bohr described electrons as only able to occupy certain energy levels / distances from nucleus. [1 mark]
4. Protons [1 mark] and neutrons [1 mark]
5. a) Grey box = alpha source [1 mark] gold square = thin gold foil [1 mark]
   b) All the alpha particles would go straight through. [1 mark]
   c) Most alpha particles went straight through [1 mark] (some were deflected / scattered at small angles) and very few (approx. 1 in 8000) bounced back / were scattered (deflected) back to the source [1 mark]
6. Marks in three bands according to level of response.

Level 3 [5–6 marks]: Clear and detailed explanation. Experiment results clearly and logically tied to correct conclusions / features of model.
Level 2 [3–4 marks]: An attempt to link conclusions / features of model to results but logic used may not be clear.
Level 1 [1–2 marks]: Explanations vague and not linked to results.
Level 0: No relevant content.
Points which should be made on the features of the Rutherford model : • Small central nucleus where most of mass is concentrated. • Nucleus is charged. • Electrons surround nucleus. • Most alpha particles went straight through, implying atom is mostly empty space. • Most alpha particles do not go near enough to the nucleus to be deflected. • Some are deflected implying the positively charged alpha particles are repelled by a positively charged 'nucleus'. • Very few bounced back, implying there is a very small, dense positive area we now call the nucleus.

Radioactive decay

1. Nuclear electron [1 mark]
2. Nucleus, unstable, random, becquerels / Bq, count rate [1 mark each]
3. a) Neutron [1 mark]
   b) Beta [1 mark]
   c) Gamma (ray) [1 mark]
   d) 2 protons, [1 mark] 2 neutrons, [1 mark] (only 1 mark for 'helium nucleus')
   e) A neutron decays into a proton and an electron. [1 mark] The electron is emitted from the nucleus. [1 mark]

Comparing alpha, beta and gamma radiation

1.

beta	**thin sheet of aluminium**	**a few metres (accept 1–3 m)**
gamma	several cm of lead	**very far**
alpha	**paper**	1–2 cm

2. a) Causes atoms to become ions [1 mark]
   b) Alpha [1 mark]
3. Beta will not penetrate body so radiation emitted by tracer cannot be detected outside body / beta very ionising so dangerous to use as tracer [1 mark]
4. a) Beta because it penetrates through outer layer of skin / alpha does not [1 mark]
   b) Best to limit penetration depth [1 mark] so fewer side effects to other organs and body tissues [1 mark]
5. Gamma would pass straight through so count rate measured on other side of paper would not change with thickness. [1 mark] Alpha absorbed by paper and air between emitter and detector, whatever the thickness of the paper. [1 mark] Beta absorbed more by thick paper than thin paper, so gives different readings for different thicknesses. [1 mark]

Radioactive decay equations

1. ${}^{4}_{2}He$ [1 mark]
2. ${}^{0}_{-1}e$ [1 mark]
3. ${}^{14}_{6}C \rightarrow {}^{14}_{7}N + {}^{0}_{-1}e$, 1 mark for 7, 1 mark for ${}^{0}_{-1}e$
4. 225 [1 mark] ${}^{4}_{2}He$ [1 mark]
5. 209 [1 mark] 83: 1 mark
6. a) Beta / nuclear electron [1 mark]
   b) ${}^{60}_{28}$ [1 mark] ${}^{0}_{-1}e$ [1 mark]

Half-lives

1. Average time it takes for activity of sample to halve [1 mark]
2. a) Attempt to read off graph from count rate 400 to 200 (or other pair of values) [1 mark] 30 +/– 2 s [1 mark]
   b) 6 times for count rate to halve [1 mark] 6 × answer to part a = 180 s [1 mark]
3. Record background count-rate, [1 mark] deduct this from each measurement [1 mark]
4. a) [2 marks for points plotted correctly] (1 mark for one plotting error, 0 if more)
   b) Smooth curve, not dot-to-dot straight lines [1 mark]
   c) 3.8 +/– 0.2 [1 mark]
   d) 3.8 +/– 0.2 [1 mark]

e) 5 half-lives to get to 6 counts/s [1 mark] working: 5 × their answer to **c)** [1 mark] 19 +/– 1 [1 mark]

5. 3 half-lives [1 mark] so 3 × 6 hours [1 mark] = 18 hours [1 mark]
6. 3 half-lives [1 mark] = 600 years in total / 3 [1 mark] 200 years [1 mark]

Radioactive contamination

1. a) Any two of: thin sheet of aluminium or equivalent between you and source, increase distance to source, reduce exposure time [1 mark each]
   b) Any two of: sheet of lead between you and source, increase distance to source, reduce exposure time [1 mark each]
2. Hard to be irradiated by alpha as it is absorbed by 1 or 2 cm air [1 mark] and by clothes or dead skin cells [1 mark]
3. Contaminated, [1 mark] because contamination involves contact with radioactive source, not just having radiation emitted pass through. [1 mark]
4. Marks in two bands according to level of response.

Level 2 [3–4 marks]: Detailed explanation with logical links between clearly identified relevant points.
Level 1 [1–2 marks]: Relevant, but separate, points are made. Logic may be unclear.
Level 0: No relevant content.
Points that should be made: • Food is safe to eat when irradiated as no radioactive material has been in contact with the food. • Radiation does not make the food radioactive. • So no radiation emitted by food when it is inside body. • If food comes into contact with radioactive source it is contaminated. • Contamination harmful as when food was ingested it would emit radiation inside body. • This is ionising, so can harm cells.

Background radiation

1. a) Any two of: rocks, soil, radon gas in air, cosmic rays, some foods [1 mark each] (1 mark for radioactive isotopes in the environment)
   b) Any two of: medical treatments / tracers / radiotherapy, exposure if work in the radiotherapy / radiography departments of hospitals or nuclear power station, nuclear weapon tests fall-out, nuclear accident fall-out, nuclear waste, waste from hospitals. [1 mark each] (Do **not** accept X-rays)
2. No, as different types of rocks in different areas / some rocks are radioactive and some are not [1 mark] so amount of radiation emitted from rocks / soils / buildings made of stone will vary. [1 mark]
3. To ensure measured count rate is just activity of sample, [1 mark] otherwise it introduces a systematic error / all readings differ from true value by a consistent amount. [1 mark]
4. Marks in two bands according to level of response.

Level 2 [3–4 marks]: Method described in order that makes sense and could be followed by someone else to obtain valid results. Method addresses the issue of improving accuracy.
Level 1 [1–2 marks]: Basic description of measurements needed with no indication of how to use them or how to improve accuracy.
Level 0: No relevant content.
Points that should be made: • Geiger–Muller tube to record number of decays / detect ionising radiation. • Reading should be taken over a long time. • So that random fluctuations in the background count over time are reduced. • Count then divided by number of seconds the reading was taken over, to give count rate / counts per second. • Take average of several readings. • To improve accuracy by reducing effect of random fluctuations. • Known sources of radiation should be removed.

Uses and hazards of nuclear radiation

1. a) Gamma radiation is less ionising, so less harmful to body, [1 mark] radiation passes through body allowing it to be detected. [1 mark]
   b) 6 hours, because gives enough time for tracer to circulate / path it takes to be monitored / measurements to be made [1 mark] but decays sufficiently quickly / not remain radioactive very long so person will not be irradiated for long. [1 mark]
2. a) Gamma radiation is only one which will penetrate deep enough into body, [1 mark] the others would be absorbed. [1 mark]
   b) One intense beam may harm healthy tissue it passes through, [1 mark] intersecting beams minimise harm to healthy tissue but combine to transfer enough energy to kill the cancerous cells. [1 mark]
3. It seems concerning that levels are so much higher than elsewhere [1 mark] but they are still comfortably below suggested limit. [1 mark]

Nuclear fission

1. Neutron [1 mark]
2. Gamma rays [1 mark]
3. Two or three neutrons [1 mark]

4. a) Neutrons released when a single **nucleus** splits (do **not** accept atom) [1 mark] absorbed by further (unstable) nuclei [1 mark] more neutrons released cause more and more uranium nuclei to split. [1 mark]
   b) In an uncontrolled chain reaction, amount of energy released keeps on increasing / need to prevent increase in reaction rate [1 mark] which would be very dangerous / rapid release energy / cause explosion. [1 mark]
   c) Neutrons [1 mark]
5. Store of (nuclear) energy in uranium fuel decreases, [1 mark] transferred to store of kinetic energy in (fast-moving) neutrons. [1 mark]

Nuclear fusion

1. Nuclei [1 mark]
2. Some of the mass of the nuclei is lost in the reaction and converted to energy of radiation. [1 mark]
3. Similarity: both involve release of energy. [1 mark] Difference: fusion involves creating a larger nucleus and fission involves creating smaller nuclei / fission requires a neutron to trigger it. [1 mark]

Section 5: Forces

Scalars and vectors

1. 10 km due east [1 mark]
2. Mass, [1 mark] time [1 mark]
3. Direction, size, displacement, scalar [1 mark each]
4. 10 m east, 10 m north, 20 m east [1 mark]

Speed and velocity

1. $\frac{100}{2.0}$ [1 mark] = 50 km/h [1 mark]
2. Velocity = 200 m/s [1 mark] direction: north [1 mark]
3. 500 s [1 mark]
4. a) 100 m [1 mark]
   b) 100/9.9 [1 mark] = 10.1 m/s [1 mark] 3 sig figs [1 mark]
5. Time taken = 3 s [2 marks]
6. a) 1.1, [1 mark] 10, [1 mark] 3.5 [1 mark]
   b) 113 000 m [1 mark]
   c) 113 000/(280 × 60) [1 mark] = 6.7 m/s [1 mark]
7. A: steady speed, B: not moving, C: getting slower [1 mark each]
8. a) Gradient = $\frac{150}{10}$ [1 mark] = 15 m/s [1 mark]
   b) Steady speed [1 mark]
9. Tangent drawn to the gradient at distance 1.5 m, [1 mark] substitution of values in $\frac{\text{change in } Y}{\text{change in } X}$, [1 mark] estimate in range of 4–6 m/s [1 mark]

Acceleration

1. $\frac{6.4 - 2.0}{4.0}$ [1 mark] = 1.1 m/s² [1 mark]
2. $\frac{25}{2.1}$ [1 mark] = 11.9(0476) [1 mark] = 12 (to 2 sig figs) [1 mark]
3. $\frac{7.7 - 6.5}{5}$ [1 mark] = 0.24 [1 mark] m/s² [1 mark]
4. 0–5 seconds: constant velocity, [1 mark] 5–10 seconds: acceleration, [1 mark] 10–15 seconds: constant velocity, [1 mark] 15–20 seconds: deceleration [1 mark]
5. $\frac{\text{change in } Y}{\text{change in } X} = \frac{6}{2}$ [1 mark] = 3 m/s² [1 mark]
6. a) $\frac{15}{3}$ [1 mark] = 5 m/s² [1 mark]
   b) $\frac{15}{2}$ [1 mark] = 7.5 m/s² [1 mark]
   c) $(\frac{1}{2} \times 3 \times 15) + (3 \times 15) + (\frac{1}{2} \times 2 \times 15)$ [1 mark]
   = 82.5 m [1 mark]
   (1 mark for marking correct area on graph)
7. Number of squares in enclosed area = 34 [1 mark] area of each square = 5 × 10 = 50 m [1 mark] distance = 1700 m [1 mark]

Equation for uniform acceleration

1. $v^2 - 0 = 2 \times 0.5 \times 80$ [1 mark] $v = \sqrt{80} = 8.9$ m/s [1 mark]
2. $v^2 - 0 = 2 \times 1.6 \times 1.3 = 4.16$ m/s, [1 mark] $v = \sqrt{4.16} =$ 2.0 m/s [1 mark]
3. $v^2 - 25 = 2 \times 0.3 \times 50$ [1 mark] $v = \sqrt{55} = 7.4$ m/s [1 mark]

Forces

1. Tension, friction, normal contact, gravitational [1 mark each]
2. One mark for each correct arrow [4 marks] Note: arrows for C and D should be the same length. Arrows for A and B should be the same length but twice the length of C and D.

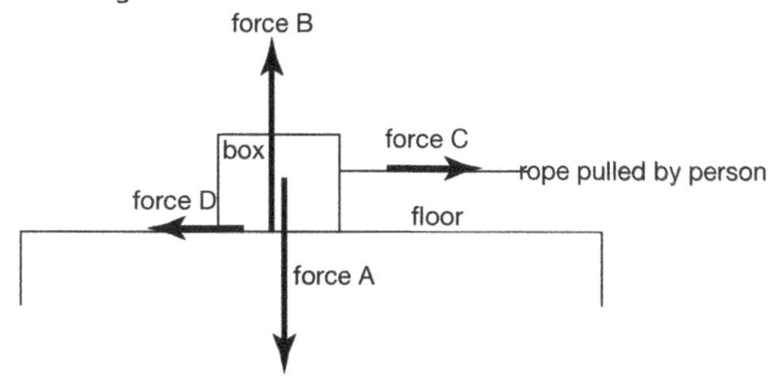

3. A = air resistance, B = gravitational force, C = normal contact force, D = normal contact force, E = engine driving force, F = friction, G = friction [1 mark each]

Moment of a force

1. 10×0.8 [1 mark] = 8 [1 mark] N m [1 mark]
2. Weight of pebble = $21 \div 16$ [1 mark] = 1.31(25) [1 mark] = 1.3 (2 sig figs) [1 mark]
3. **a)** 1×20 [1 mark] = 20 [1 mark]
   b) $2.8 \times d = 20 \times 1$ [1 mark] $d = 7.1(4)$ cm [1 mark] = 7.1 cm (2 sig figs) [1 mark]
   c) 32 cm [2 marks]

Levers and gears

1. **a)** 800×0.5 [1 mark] = 400 [1 mark] N m [1 mark]
   b) $400 = F \times 1.5$ [1 mark] $F = 267$ N [1 mark]
2. Small, large, small [1 mark each]
3. **a)** 200×0.4 [1 mark] = 80 N m [1 mark]
   b) Anticlockwise [1 mark]
   c) Force $\times$ 0.80 = 80 [1 mark] Force = $80 \div 0.8$ = 100 N [1 mark]

Pressure in a fluid

1. $\frac{10}{0.018}$ [1 mark] = 555.6 Pa [1 mark] = 556 Pa (3 sig figs) [1 mark]
2. **a)** 0.45 m^2 [1 mark]
   b) $1\,100\,000 \times 0.45$ [1 mark] = 495 000 [1 mark] unit: N [1 mark]
3. $100\,000 \times 1.7$ [1 mark] = 170 000 N [1 mark]
4. **a)** $15.0 \times 1030 \times 9.8$ [1 mark] = 151 410 = 151 000 Pa [1 mark]
   b) 252 000 Pa [1 mark]
   c) $10\,900 \times 1030 \times 9.8$ [1 mark] = 110 000 246 = 110 000 000 Pa [1 mark]
   d) 1090 times [1 mark]

Atmospheric pressure

1. **a)** 29 000 Pa [1 mark] $29\,000 \times 1.7$ [1 mark] = 49 000 N [1 mark] (accept 48 000 – 51 000 N)
   b) Molecules of gas in air collide with surface, [1 mark] exert force (when they collide) with surface. [1 mark]
   c) More molecules collide with surface every second [1 mark] so larger force is exerted on surface. [1 mark]
   d) At sea level there is more air above us than at the top of Mount Everest. [1 mark] The molecules of gas in the air at the top of Everest are not as squashed together as they are at sea level. [1 mark]
2. The 60 mile thickness of the atmosphere is small compared with the radius of the Earth. [1 mark]
3. **a)** [1 mark for arrow]

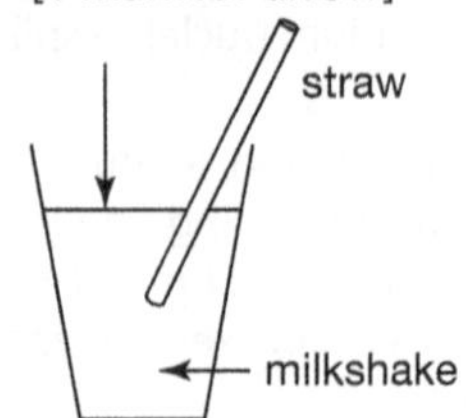

   b) Sucking on straw reduces air pressure inside straw. [1 mark]
   Atmospheric pressure is now greater than air pressure inside straw. [1 mark] Atmospheric pressure pushes milkshake up straw. [1 mark]

Gravity and weight

1. Doesn't change [1 mark] decreases [1 mark]
2. **a)** 900×9.8 [1 mark] = 8820 [1 mark] N [1 mark]
   b) 900×3.7 [1 mark] = 3330 [1 mark] N [1 mark]
3. Multiplying mass by a constant factor, such as 2 [1 mark] causes weight to also be multiplied by 2 [1 mark]
4. X at centre, [1 mark] arrow downwards from X [1 mark]

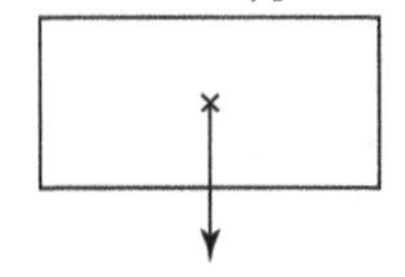

5. **a)** Newtonmeter [1 mark]
   b) $2.7 \div 9.8$ [1 mark] = 0.28 N [1 mark]
6. Reduction in mass = 1.7 kg, [1 mark] reduction in weight = 1.7×9.8 [1 mark] = 16.66 [1 mark] = 17 (to 2 sig figs) [1 mark]

Resultant forces and Newton's 1st law

1. 6 N [1 mark] direction: left [1 mark]
2. **a)** If resultant force on moving object is zero [1 mark] object continues to move at same speed and in same direction. [1 mark]
   b) 0 N [1 mark]

3. [1 mark each for correctly labelled forces] [1 mark for arrows same length]

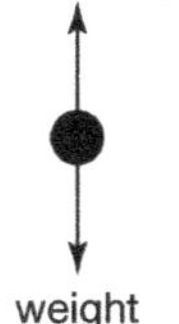

4. a) [1 mark each for correct length and position of a component]

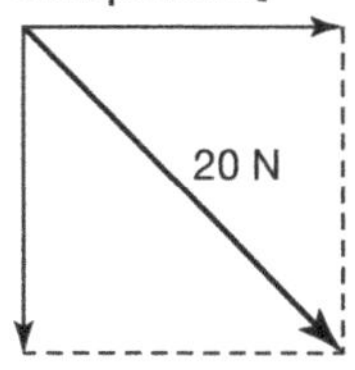

b) Horizontal = 14 N, [1 mark] vertical = 14 N [1 mark]

Forces and acceleration

1. a) Air track removes friction [1 mark]
   b) Air resistance / drag [1 mark]
   c) $\frac{0.10}{0.77}$ [1 mark] = 0.13 m/s [1 mark]
   d) $\frac{0.10}{0.30}$ [1 mark] = 0.33(3) [1 mark] = 0.33 (2 sig figs) [1 mark]
   e) $(0.33)^2 - (0.13)^2 = 2 \times$ acceleration $\times 0.5$ [1 mark] acceleration = 0.092 m/s^2 [1 mark]
   f) Repeat the anomalous measurement. [1 mark] If same result is obtained, either repeat all measurements [1 mark] or include anomalous point when drawing the best fit line. [1 mark]
   g) Acceleration is directly proportional to force. [1 mark] Best fit line is straight and passes through origin. [1 mark]
2. a) 2800 N [1 mark]
   b) 1.4 [1 mark] m/s^2 [1 mark]
3. a) Thrust: upwards, air resistance: downwards, weight: downwards [1 mark each]
   b) 2 100 000 N 1 mark]
   c) 2 100 000 ÷ 550 000 [1 mark] = 3.8 m/s^2 [1 mark]
4. a) [1 mark for each correctly labelled arrow] [1 mark if weight arrow longer than air resistance arrow]

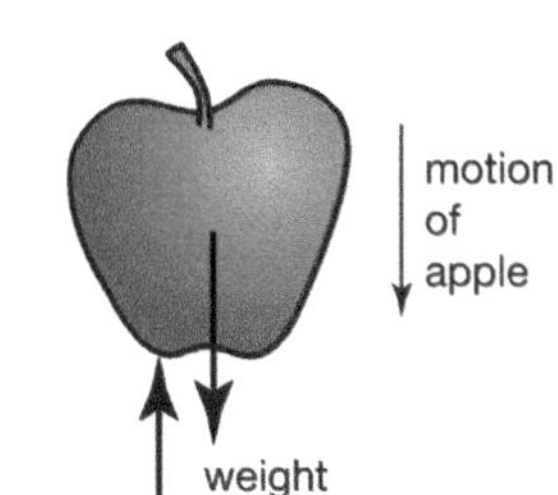

   b) 0.10 kg [1 mark]
   c) 0.98 N [1 mark]
   d) 0.68 N [1 mark]
   e) $a = \frac{0.68}{0.1}$ = [1 mark] = 6.8 m/s^2 [1 mark]

Terminal velocity

1 a) A [1 mark]
   b) A: increasing, B: increasing, C: not changing [1 mark each]
   c) B: decreasing, C: zero [1 mark each]
   d) B: decreasing, C: zero [1 mark each]
2. Marks in three bands according to level of response.

Level 3 [5–6 marks]: Clear coherent description of each stage of the motion in sequence, considering velocity and acceleration explained in terms of the relative size of air resistance and weight.
Level 2 [3–4 marks]: Clear description in sequence of some of the stages of the motion with some explanation with reference to the relative size of air resistance and weight. Some detail of either velocity or acceleration may be missing and the stages of the explanation may not be in an order that makes sense.
Level 1 [1–2 marks]: Some correct comments regarding velocity and acceleration but does not include any explanation.
Level 0: No relevant content.
Points that should be made: • Acceleration greatest as she leaves aircraft. • On leaving aircraft, air resistance is much smaller than her weight. • Her acceleration decreases as air resistance gets bigger. • Her velocity increases until she reaches her terminal velocity, before opening her parachute. • At her (first) terminal velocity, air resistance is equal to her weight. • Rapid deceleration on opening her parachute. • Parachute creates a big increase in air resistance. • As she decelerates, the air resistance is bigger than her weight. • Her 2nd terminal velocity is much lower than the first. • At her 2nd terminal velocity, air resistance is (again) equal to her weight.

Newton's third law

1. a) Opposite direction [1 mark]
   b) Same size [1 mark]
   c) Same type [1 mark]
2. B, A [1 mark]
3. a) Gravitational force [1 mark]
   b) Normal contact force [1 mark]
   c) Gravitational, upwards, the Earth [1 mark each]
   d) Normal contact, downwards, the floor / Earth [1 mark each]
4. Combustion chamber exerts downward force on exhaust gases, [1 mark] exhaust gases exert upward force on combustion chamber of the rocket. [1 mark]

Work done and energy transfer

1. a) 15×10 [1 mark] = 150 [1 mark] J (accept N m) [1 mark]
   b) Friction, [1 mark] opposite to direction trolley is moving [1 mark]
   c) Chemical energy store in shopper's body decreases, [1 mark] thermal energy store of surroundings increases [1 mark]
2. a) 1000×15 [1 mark] = 15 000 J [1 mark]
   b) $200 \times 9.8 \times 6.5$ [1 mark] = 12 740 = 13 000 J [1 mark]
   c) Some energy is dissipated as thermal energy [1 mark]
3. a) 800×100 [1 mark] = 80 000 J [1 mark]
   b) 80 000 W [1 mark]

Stopping distance

1. a) Kinetic energy = $\frac{1}{2}mv^2$ [1 mark] $0.5 \times 800 \times 15^2$ [1 mark] = 90 000 J [1 mark]
   b) 15×0.60 [1 mark] 9.0 m (accept 9) [1 mark]
   c) $5000 \times d = 90\,000$ [1 mark] $d = 18$ m [1 mark] (allow error carried forward)
   d) 27 m [1 mark] (allow error carried forward)
2. a) $0.5 \times 800 \times 30^2$ [1 mark] = 360 000 J [1 mark]
   b) 30×0.60 [1 mark] = 18 m [1 mark]
   c) $5000 \times d = 360\,000$ [1 mark] $d = 72$ m [1 mark]
   d) 90 m [1 mark]
   e) If speed of a car is doubled then distance that car travels before it can stop is 3.3 times bigger. [1 mark]
3. Any two from: tiredness, distracted by something, has drunk alcohol, has taken drugs [1 mark each]
4. a) $13 + 30 = 43$ m, [1 mark] lines drawn on graph to get thinking distance [1 mark]
   b) Any two from: wet road, icy road, bald tyres, brakes in poor condition [1 mark each]
5. a) Transferred to thermal energy store in brakes. [1 mark]
   b) Temperature increases [1 mark]
   c) Brake temperature rises more rapidly [1 mark] which may cause them to overheat and be less effective. [1 mark]
   d) Seat belt [1 mark]

Force and extension

1. a) Clamp, [1 mark] weight [1 mark]
   b) 0.05 m [1 mark]
   c) $\frac{1}{2} \times 35 \times 0.05^2$ [1 mark] = 0.044 J [1 mark] 2 sig figs [1 mark]
2. a) 20, 30, 10, 40 [1 mark each]
   b) 3 [1 mark]
   c) 4 [1 mark]
3. a) 2.3, 4.5, 6.8, 9.1, 11.4, 13.7 [2 marks]
   b) To prevent the stand from toppling over. [1 mark]
   c) Extension is directly proportional to stretching force. [1 mark]
   d) Graph is a straight line [1 mark] through the origin. [1 mark]
   e) Weight in range 2.7–2.8 N, [1 mark] one line drawn on graph from 0.11 m to best fit line and 2nd line vertically downwards to stretching force axis. [1 mark]

Momentum

1. kg × m/s = kg m/s [1 mark]
2. a) Vector [1 mark]
   b) Momentum = mass × velocity, [1 mark] velocity is a vector [1 mark]
3. 800×20 [1 mark] = 16 000 [1 mark] kg m/s [1 mark]
4. Convert 100 km/h to 27.8 m/s [1 mark] 0.057×27.8 [1 mark] = 1.6 kg m/s [1 mark] 2 sig figs [1 mark]
5. 4800 kg m/s, 6400 kg m/s, 4000 kg m/s [1 mark each]
6. $1200 \div 800$ [1 mark] 1.5 m/s [1 mark]

Conservation of momentum

1. Total momentum before event is equal to total momentum after event, or (total) momentum stays the same [1 mark]
2. a) 200 kg m/s [2 marks]
   b) 120 kg m/s [2 marks]
   c) 200 kg m/s [1 mark]
   d) 80 kg m/s [1 mark]
   e) 0.8 m/s [2 marks]
3. Because the rugby players are running in opposite directions, one has positive momentum and the other has negative momentum. [1 mark] If their value for mass times velocity is the same then their momentum can add up to zero. [1 mark]

Rate of change of momentum

1. Larger, larger, larger, smaller, smaller [1 mark each]
2. a) $\frac{40 \times 4}{0.2}$ [1 mark] = 800 N [1 mark]
   b) Impact force is much bigger (without crash mat), [1 mark] momentum changes much more quickly (without crash mat). [1 mark]
3. 'Time taken' is the denominator [1 mark] so a bigger value for 'time taken' means that $\frac{\text{momentum change}}{\text{time taken}}$ is smaller [1 mark] and, since this equals the impact force, force is also smaller. [1 mark]
4. Seat belt increases time taken to stop, meaning momentum change occurs at a slower rate, [1 mark] reducing impact force. [1 mark]

Section 6: Waves

Transverse and longitudinal waves

1. Waves transfer energy but no matter. [1 mark]
2. Perpendicular, [1 mark] parallel [1 mark]
3. a) Water waves / electromagnetic waves [1 mark]
   b) Sound waves [1 mark]
4. a) Longitudinal [1 mark]
   b) Correctly labelled where coils most bunched [1 mark]
   c) Correctly labelled where coils furthest apart [1 mark]
5. a) Vertically, up and down [1 mark] (Allow particles at surface move in a circle **if** also state that there is no overall horizontal movement.)
   b) Wave moves water (and boat) up and down only, [1 mark] wave travels outwards (from source) but no water travels outwards [1 mark]

Frequency and period

1. 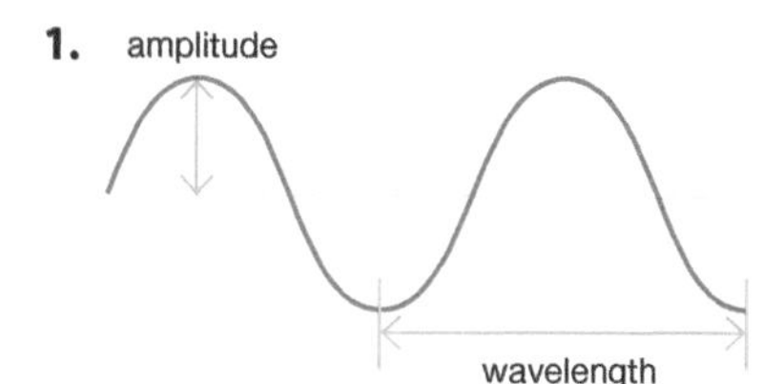

   a) [1 mark if amplitude labelled correctly]
   b) [1 mark if wavelength labelled correctly]
2. a) Number of complete waves produced / passing a point each second. [1 mark] hertz / Hz [1 mark]
   b) Distance between two points on a wave and equivalent point on the adjacent wave moving in phase. [1 mark] metres / m [1 mark]
   c) Time for one complete wave to pass a point [1 mark] (allow reciprocal or inverse of frequency), seconds / s [1 mark]
3. a) 1/10 [1 mark] = 0.1 s [1 mark]
   b) Rearrange formula [1 mark] 1/0.5 [1 mark] = 2 Hz [1 mark]
   c) 1/5000 [1 mark] = 0.0002 s [1 mark]

Wave speed

1. Medium [1 mark]
2. a) Wave speed = frequency × wavelength [1 mark]
   b) m/s / metres per second, [1 mark] Hz / hertz, [1 mark] m / metres [1 mark]
3. a) 1 m, [1 mark] 4 m [1 mark]
   b) Rearrangement [1 mark] 5/4 [1 mark] = 1.25 Hz [1 mark]
4. a) 10/60 per second [1 mark] = 0.167 Hz [1 mark]
   b) 0.167 × 2 [1 mark] = 0.334 m/s [1 mark]
5. 100/600 [1 mark] = 0.166(6) [1 mark] = 0.17 Hz (2 sig figs) [1 mark]
6. Marks in three bands according to level of response.

Level 3 [5–6 marks]: Method described in detail including equation required to calculate speed. Steps in an order that makes sense and could be followed by someone else to obtain valid results. Some valid suggestions about reducing errors made, with reasons.
Level 2 [3–4 marks]: Method described clearly but some detail may be missing and steps may not be in an order that makes sense. At least one valid suggestion about reducing errors made, with reasons.
Level 1 [1–2 marks]: Basic description of measurements needed with no indication of how to use them or reduce errors.
Level 0: No relevant content.
Points that should be made: • Example for how to produce a loud noise – e.g. bursting a balloon or banging wooden blocks together.
• Time and distance measured. • Suitable equipment chosen to measure these. • Calculate speed using equation speed = distance/time • Allowance for distance sound travels = 2 × distance of person from wall / reflecting surface • Do repeats and calculate a mean. • Reference to problems with reaction time with measurements. • Improve method by either changing equipment e.g. using microphones to automatically detect the sound, or by improving reaction time issue (e.g. averaging multiple times).

7. a) Count number of waves produced in set time (e.g. 1 minute), [1 mark] **or** divide number of waves produced in one minute by 60 [2 marks]
   b) Waves constantly moving, [1 mark] alternative offered e.g. take photo of waves with a ruler next to it [1 mark]
   c) Describe a calculation from wavelength and frequency instead of a measurement [1 mark only] **or** measure time it takes for a wave to travel a set distance (e.g. length of ripple tank) [1 mark] calculate distance/time to work out speed [1 mark]

Reflection and refraction of waves

1. a) Angle of reflection should be (when judged by eye) equal to the angle of incidence. [1 mark]
   b) Angle between the reflected ray and the dotted line [1 mark]

c) Use a ray box and slit to produce narrow ray / beam of light, [1 mark] mark (on paper) ray of light reflected from mirror, [1 mark] measure angle of reflection with a protractor [1 mark]

d) Width of light ray makes it difficult to judge where centre of the ray is / mark path precisely [1 mark]

2. a) Within block: straight line at smaller angle to normal than angle of incidence, [1 mark] ray leaving block: straight line at larger angle to normal than angle of incidence, [1 mark] exiting ray parallel to original ray [1 mark]

b) All light absorbed [1 mark]

3. E.g. aluminium foil after being scrunched [1 mark]

4. E.g. mirror or very smooth water surface [1 mark]

Sound waves

1. Longitudinal, [1 mark] 20, [1 mark] 20 000 [1 mark]
2. 290 × 1.2 [1 mark] = 348 (m/s) [1 mark]
3. Solids [1 mark] particles closer together (in a solid) so pass on vibrations / energy more effectively [1 mark]
4. Vibrations (accept kinetic energy of particles) in air make ear drum vibrate [1 mark] passing on vibrations (accept transfers energy) to other parts of ear sending signal to brain [1 mark]
5. Magnetic field (around coil), [1 mark] interacts with this magnetic field, [1 mark] change in the force, [1 mark] vibrates and so emits sound waves [1 mark]
6. a) Cone vibrates at higher rate / more quickly [1 mark]

   b) Cone vibrates with greater amplitude [1 mark]

Ultrasound and echo sounding

1. Speeds, absorbed, reflected, [higher [1 mark each]
2. Marks in two bands according to level of response.

Level 2 (3–4 marks): Detailed explanation of measurements needed, including how to apply them to a distance calculation
Level 1 (1–2 marks): Basic description of measurements needed with no indication of how to use them.
Level 0: No relevant response.

Points which should be included:
- Ultrasound transmitter and receiver both placed on mother's tummy.
- Pulses of ultrasound sent through skin into womb.
- Ultrasound is (partially) reflected when it meets a boundary between two different media.
- Travel at different speeds through different media.
- Know speed of ultrasound through tissues in baby's body.
- Detect time delay between emitting and receiving ultrasound signal.
- Distance = speed × time
- Distance then divided by two as signal travels twice the distance to target.
- Reflections from back and front of head (or an organ) return at different times (from different depths).
- Size of structure calculated from separate echoes.

3. Any 2 non-medical uses, e.g. detect fault in pipe, depth to a pipe, echo sounding, detect objects in deep water, measure water depth [1 mark each]
4. Distance = speed × time [1 mark] = 330 × 2.5/2 [1 mark] = 412.5 m [1 mark] (allow 2 marks for 825 m with working shown)

Seismic waves

1. Longitudinal, faster, transverse, liquids [1 mark each]
2. a) Parallel to direction of wave / energy transfer [1 mark]

   b) Perpendicular to direction of wave / energy transfer [1 mark]
3. a) P-waves travel faster than S-waves [1 mark]

   b) At least part of the core is liquid [1 mark] as no S-waves travel through it and S-waves cannot travel through liquid. [1 mark]

   c) The sudden change in P-wave velocity is due to a different state in the Earth's structure, e.g. travelling from solid to liquid. [1 mark]

The electromagnetic spectrum

1. Transverse, frequencies, highest, lowest [1 mark each]
2. Speed [1 mark]
3. a) X-rays / gamma rays [1 mark]

   b) Radio waves [1 mark]

   c) Red [1 mark]
4. Speed is constant [1 mark] so to keep speed constant, the higher the wavelength the lower the frequency [1 mark]
5. Frequency = wave speed/wavelength, [1 mark] 300 000 000/0.000 000 01 [1 mark] = 3×10^{16} Hz [1 mark]

Reflection, refraction and wave fronts

1. Wave fronts all straight lines perpendicular to ray, [1 mark] angle of incidence roughly equal to angle of reflection [1 mark]
2. a) Minimum 2 wave fronts shown: wave fronts entering block change direction so bend towards normal, [1 mark] wave fronts closer together, [1 mark] wave fronts leaving block change direction to bend away from normal, [1 mark] unchanged wavelength (by eye) compared to original wave fronts [1 mark]
   b) Change of speed when a wave travels from one medium to a different medium [1 mark]
3. Longer / increases, [1 mark] no change [1 mark]
4. Wave speed = frequency × wavelength. [1 mark] Wave speed gets higher and frequency doesn't change [1 mark] so wavelength increases. [1 mark]

Emission and absorption of infrared radiation

1. All objects absorb and emit infrared radiation. [1 mark]
2. Both the temperature and surface of the object [1 mark]
3. a) Same volume of water in each [1 mark] same starting temperature [1 mark]
   b) Shiny silver [1 mark] because worst emitter of infrared radiation / reflects infrared radiation back inside. [1 mark]
   c) Matt black [1 mark] because it is best emitter of infrared radiation / radiates infrared radiation away from bottle. [1 mark]
4. a) Matt black [1 mark]
   b) Thermometer [1 mark]
   c) Any one of: infrared sensor has better resolution, better thermal contact, removes random reading / measurement error, records data automatically [1 mark]
5. Temperature is increasing [1 mark] because it takes in more energy than it gives out. [1 mark]

Uses and hazards of the electromagnetic spectrum

1. Radio waves: television and radio
   Microwaves: Satellite communications, cooking food
   Infrared: electric heaters, cooking food, night-vision cameras
   Visible light: fibre optic communications
   Ultraviolet: energy efficient lamps, sun tanning
   X-rays and gamma rays: medical imaging and treatments [1 mark each]
2. Absorbed by bone [1 mark]
3. Ultraviolet [1 mark]
4. Infrared [1 mark]
5. UV, X-rays and gamma, [1 mark] most ionising [1 mark]
6. a) Wavelength (accept frequency) [1 mark]
   b) Absorbed (by the atmosphere) [1 mark]
7. a) Microwaves not refracted / reflected by ionosphere but radio waves are [1 mark] (Accept microwaves can pass through ionosphere but radio waves cannot, unless transmitted at very large angle of incidence.)
   b) Smoke absorbs visible light, [1 mark] infrared is transmitted through / not absorbed by smoke. [1 mark]
8. a) 100 mSv [1 mark]
   b) No, because a single X-ray is well below safe radiation dose for 5 years. [1 mark]
   c) 100/1.5 [1 mark] = 66 [1 mark] If they do several a day they may exceed that limit over 5 years. [1 mark]
   d) Leave room X-ray machine whilst image is taken. [1 mark]

Radio waves

1. Radio / TV broadcasting, [1 mark] communications network [1 mark] (do **not** accept satellite communications)
2. Low, low, longer [1 mark each]
3. Oscillating / changing current (in electric circuit), [1 mark] produces electromagnetic wave [1 mark]
4. Radio waves absorbed, [1 mark] causes charges in receiver circuit to oscillate / creates an alternating current [1 mark]
5. Frequency [1 mark]
6. Ions [1 mark] because charged (and radio waves are produced by oscillating charges) [1 mark]

Colour

1. Transverse, [1 mark] travel same speed [1 mark]
2. Frequency, [1 mark] wavelength [1 mark]
3. Blue filter absorbs other colours in white light, [1 mark] only transmits blue. [1 mark]
4. a) T-shirt absorbs other colours [1 mark] only reflects red. [1 mark]
   b) Black, because blue shirt only reflects blue and only red is incident on it. [1 mark]

Lenses

1. Refraction, thicker, thinner, virtual [1 mark each]
2. Level 2 [3–4 marks]: Good, labelled diagram clearly showing focal length with good attention to detail with symbols, straight rays of light and precise labelling.

Level 1 [1–2 marks]: Attempt at a relevant diagram but diagram may be only partially labelled or care may not have been taken to show focal length precisely.
Level 0: no relevant diagram.
Points which should be included: • Diagram showing parallel rays coming to a focus after lens. • Focal length explained and identified as distance from centre of lens to principal focus. • Correct symbol for a convex lens. • All rays should be straight and not dotted. • Arrows in correct direction.

3. a) 6 mm [1 mark]
   b) 6/0.3 [1 mark] = 20 [1 mark]
4. Horizontal line from top of object to lens, then straight line from there through F, [1 mark] straight line from top of object through middle of lens and out other side, [1 mark] image arrow drawn at intersection [1 mark]

A perfect black body

1. Absorbs, [1 mark] emitter [1 mark]
2. Cooling down / decreasing [1 mark] because losing more energy than it is gaining. [1 mark]
3. No, [1 mark] it emits and absorbs energy at same rate [1 mark]
4. Different intensity, [1 mark] different / range of wavelengths [1 mark]

Temperature of the Earth

1. a) Reflected by clouds, [1 mark] absorbed (by gases in the atmosphere) [1 mark]
   b) Reflected (by surface of Earth), [1 mark] absorbed (by surface of Earth) [1 mark]
2. Certain gases (e.g. carbon dioxide and methane) in atmosphere [1 mark] absorb longer wavelength infrared radiation more strongly than short-wavelength infrared radiation [1 mark] (Allow for 1 mark, longer wavelength / radiated IR cannot pass through atmosphere / is trapped.)
3. a) Any one of: increase rate of absorption of radiation, increase rate of emission (of solar radiation), decrease rate incoming solar radiation is reflected to space [1 mark]
   b) Depends on answer to part b: if increase absorption rate of radiation → increase concentration of greenhouse gases in atmosphere. If increase rate of emission, → any one of: changes in the Sun's brightness, variations in the shape of Earth's orbit. If decrease rate of reflection of IR into space → decrease reflectivity / albedo of Earth's surface and / or clouds so less solar radiation reflected away from Earth. [1 mark]

4.

Level 2 (3–4 marks): Well-labelled diagram with correct reasons.
Level 1 [1–2 marks]: Attempt at relevant diagram including mostly correct reasons. Some arrows may point in wrong direction.
Level 0: No relevant response.
Points which should be included: • No solar radiation coming in at night. • Some emission of infrared radiation from the Earth's surface shown in a diagram. • IR radiation given out by Earth has different (longer) wavelength from IR received from the Sun • because it is cooler. • Reflection of (longer wavelength) IR off clouds back down to Earth. • Absorption of (longer wavelength) IR in atmosphere. • Description of the atmosphere trapping thermal energy due to gases that absorb some longer wavelength IR. • Greatly reduced rate of emission (of solar radiation), but slowly decreasing rate of absorption (in atmosphere) prevents temperature of Earth's surface and atmosphere falling too quickly.

Section 7: Magnetism and electromagnetism

Magnets and magnetic forces

1. Repel [1 mark] attract [1 mark]
2. Nickel [1 mark] cobalt [1 mark]
3. Near poles [1 mark] (accept closest to magnet)
4. Non-contact [1 mark] Force acts between two magnets or a magnet and a magnetic material without them needing to touch. [1 mark]
5. Permanent magnet has its own field, [1 mark] induced magnet only becomes magnetic when in a magnetic field, [1 mark] induced magnet loses magnetism when removed from field, [1 mark] induced magnet attracted to either pole of permanent magnet. [1 mark]

Magnetic fields

1. Region around magnet in which a force acts on other magnets and magnetic materials. [1 mark]
2. Steel, [1 mark] the south pole of another magnet [1 mark]
3. A pole, [1 mark] this is where field lines are closest together and force is strongest [1 mark]
4. South [1 mark]
5. Magnetic field of Earth weaker than field close to the permanent magnet, [1 mark] direction of field around permanent magnet at that point different to direction of Earth's magnetic field. [1 mark]

6. Place compass near magnet, mark dot / cross on paper / card where compass is pointing, [1 mark] after several repeats a line is drawn to join dots, [1 mark] add arrows on lines in direction compass needle points. [1 mark]

The magnetic effect of a current

1. Connect battery to either end of wire, [1 mark] start with compass far from wire, [1 mark] then show its change of direction as it gets closer to wire. [1 mark] Also accept place compass near wire, [1 mark] check direction of compass for no current, [1 mark] connect battery to wire and see if needle changes direction. [1 mark]
2. **a)** Two concentric circles [1 mark]; arrows shown clockwise on both circles [1 mark].
   b) Reverses/becomes anticlockwise [1 mark]
3. **a)** (Soft) iron [1 mark]
   b) Field pattern outside coil similar to field of bar magnet, [1 mark] minimum of 4 field lines, no field lines crossing, evenly spaced straight lines inside coil, [1 mark] correct direction of arrows from N pole to S pole [1 mark] (no mark if direction of any arrows contradict)
4. Current supply shown [1 mark] coil shown [1 mark] and iron core [1 mark]
5. Marks in two bands according to level of response.

Level 2 [3–4 marks]: Clear, coherent description and explanation of each stage of operation of the bell.
Level 1 [1–2 marks]: Description in sequence of some stages of operation of bell, some explanation though some parts may be incorrect / missing.
Level 0: No relevant response.
Points which should be made: • When switch is pressed, current passes through coil. • When current passes a magnetic field is generated around electromagnet. • This attracts iron strip. • This breaks circuit, meaning current no longer passes. • So no longer a magnetic field around the electromagnet. • So spring pulls iron strip back again. • This means circuit is once more complete and current flows again. • This cyclical process continues (as long as switch is pressed), meaning that hammer is constantly hitting the gong and generating a ringing sound.

Fleming's left hand rule

1. **a)** Conductor moves [1 mark]
   b) Conductor carrying current produces magnetic field (around wire), [1 mark] the two magnetic fields interact to produce force on conductor. [1 mark]
   c) Motor effect [1 mark] (accept catapult effect)
2. **a)** Arrow points away from centre of magnet / switch at right angles to both current and uniform field [1 mark]
   b) Stronger field [1 mark] higher current (accept increase p.d.) [1 mark]
3. $0.003 \times 0.5 \times 2$ [1 mark] $= 0.003$ N [1 mark]
4. $0.5/(0.1 \times 0.1)$ [1 mark] $= 50$ N [1 mark]

Electric motors

1. **a)** From N pole to S pole (left to right) [1 mark]
   b) Up [1 mark]
   c) Down [1 mark]
   d) Clockwise [1 mark]
   e) Forces (on each side of coil) must reverse in direction. [1 mark]
   f) Current is reversed [1 mark] using split-ring commutator (to reverse current direction every half turn) [1 mark]
   g) Higher current, [1 mark] stronger field of permanent magnet, [1 mark] more turns on coil [1 mark]

Loudspeakers

1. They both use direct current [1 mark]
2. **a)** Right [1 mark]
   b) Out of page [1 mark]
   c) Right [1 mark]
   d) Direction of current in top and bottom of coil is opposite [1 mark] but direction of force on cone must be same at top and bottom, [1 mark] achieved by reversing field. [1 mark]
   e) When current reverses, direction of force also reverses, [1 mark] so cone moves alternately in and out. [1 mark]
   f) Generates longitudinal pressure wave in air [1 mark] = sound wave [1 mark]

Induced potential

1. p.d., generator, at right angles, current, magnetic field [1 mark each]
2. **a)** Higher current [1 mark]
   b) Current in opposite direction [1 mark]
   c) No current [1 mark]
   d) Higher current [1 mark]
3. **a)** No p.d. induced [1 mark] because no relative motion / no field lines cut [1 mark]
   b) No difference in the p.d. produced, [1 mark] relative motion between magnet and wire [1 mark]
   c) Induced p.d. generates magnetic field [1 mark] opposing / in opposite direction to original change [1 mark]

Uses of the generator effect

1. Alternator / generator [1 mark]
2. Dynamo [1 mark]
3. Coil spins faster creating higher frequency a.c., [1 mark] and greater current due to greater force [1 mark]
4. a)

Level 2: [3–4 marks]: Detailed explanation with logical links between clearly identified relevant points.
Level 1: [1–2 marks]: Relevant, but separate, points are made. The logic may be unclear.
Level 0: No relevant response.
Points which should be made: • Force is applied to coil. • Movement of coil at right angles to magnetic field. • Coil cuts through field lines. • This induces a p.d. between ends of coil. • Coil is part of a complete circuit. • So the induced p.d. generates current because there is a split connection between rotating coil and external circuit OR each side of coil connects with a different side of the circuit every half turn.

   b) Line either **only** positive or **only** negative values [1 mark]
5. Line drawn demonstrates oscillations in p.d. – not just a straight horizontal line, [1 mark] oscillations symmetrical and approximately correct in shape [1 mark]

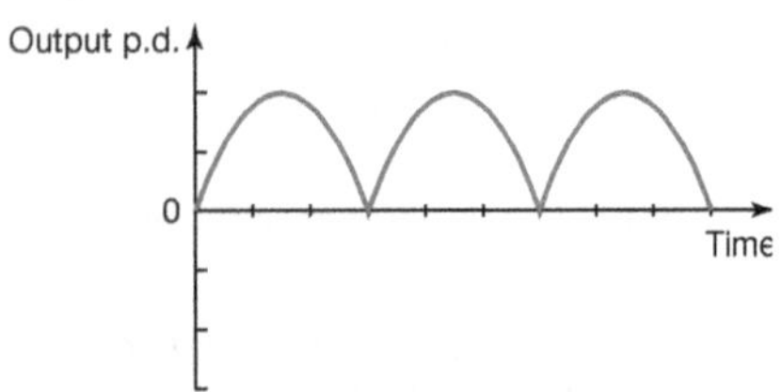

Microphones

1. a) Moving diaphragm makes coil vibrate backwards and forwards relative to (fixed) magnet, [1 mark] p.d. induced across ends of coil [1 mark]
   b) Vibration [1 mark]
2. a) Increase [1 mark]
   b) Increase [1 mark]
   c) Increase [1 mark]
3. Any two of: quieter sounds produce smaller amplitude of vibration so diaphragm less taut / less mass (to allow it to move more), coil could have more turns, or magnetic field strength of permanent magnet could be greater. [1 mark each]

Transformers

1. Decreasing the current through the cables reduces the energy losses in the cables [1 mark]
2. a) a.c. [1 mark]
   b) a.c. (in primary) induces alternating magnetic field [1 mark] in (iron) core, [1 mark] this magnetic field links with secondary coil [1 mark] which induces (alternating) voltage / p.d. across secondary (coil) [1 mark]
   c) To strengthen magnetic field [1 mark]
   d) Iron [1 mark]
3. a) $10/V_s = 5/50$ [1 mark] $V_s = 10 \times 50/5$ [1 mark] = 100 (V) [1 mark]
   b) Step-up [1 mark]
   c) $100Is = 10 \times 2$, [1 mark] $s = (2 \times 10)/100$ [1 mark] = 0.02 A [1 mark]

Section 8: Space physics

Our solar system

1. Star, planet, planet, galaxy [1 mark each]
2. It orbits Mars / the planet [1 mark]
3. From a cloud of dust and gas (nebula) [1 mark] pulled together by gravitational attraction [1 mark]
4. Mass of dust and gas cloud it forms from [1 mark]
5. a) Gravitational – inwards [1 mark] and expansion / pressure outwards [1 mark]
   b) Forces balanced / in equilibrium [1 mark]

The life cycle of a star

1. (Nuclear) fusion [1 mark]
2. Runs out of fuel / hydrogen to fuse / not enough energy to continue to fuse larger elements [1 mark]
3. Elements formed by fusion in stars [1 mark] = joining together of small nuclei to make larger ones. [1 mark] Elements heavier than iron were generated in supernovae. [1 mark]
4. No, [1 mark] our Sun too small a star to result in supernova. [1 mark]
5. All formed from dust and gas clouds, [1 mark] all form protostars, [1 mark] all enter main sequence stage. [1 mark]
6. High-mass stars form red super giants whereas low-mass stars form red giants, [1 mark] high-mass stars end with huge explosions of their outer layers / supernovae, whereas smaller mass stars do not have supernovae, [1 mark] high-mass stars go on to form black holes or neutron stars whereas low-mass stars form white / black dwarfs. [1 mark]

Orbital motion, natural and artificial satellites

1. Gravity / gravitational attraction [1 mark]
2. Natural satellites not put there by humans, whereas artificial ones were. [1 mark] Natural satellite: Moon around Earth or description of moon of another planet. [1 mark] Artificial satellite: weather satellite / spying / GPS / communications [1 mark]
3. Planet orbits Sun whereas a moon orbits a planet. [1 mark]
4. Earth travels at a steady speed round Sun [1 mark] but as its direction of motion is constantly changing so does its velocity. [1 mark]
5. **a)** Speed [1 mark]
   b) Orbital radius changes [1 mark] to maintain stable orbit [1 mark]

Red-shift

1. **a)** Light from distant galaxies has longer wavelength / higher frequency / moves towards red end of spectrum [1 mark]
   b) Red-shift [1 mark]
   c) More distant galaxies have larger increase in wavelength / larger red-shift [1 mark]
   d) They were closer together [1 mark]
2. **a)** Universe began from very small, hot, dense region, [1 mark] (massive) explosion / rapid expansion sent all matter outwards / caused Universe to expand [1 mark]
   b) Red-shift shows galaxies are moving away from each other / the Earth. [1 mark] More distant galaxies show bigger red-shift / bigger change in wavelength / frequency. [1 mark] Suggests whole Universe is expanding outwards from a small initial point / space is expanding / everything is moving away from everything else. [1 mark]
3. **a)** Supernovae [1 mark]
   b) Rate of expansion increasing [1 mark]

Dark matter and dark energy

1. **a)** A mysterious force has pushed galaxies apart, making the expansion speed up. [1 mark]
   b) Dark energy [1 mark]
2. **a)** Dark matter [1 mark]
   b) Gravitational [1 mark]
3. Mass × 4/100 = 10^{53} kg [1 mark] mass = 10^{53} kg × 25 [1 mark] = 10^{54} kg (to nearest order magnitude) [1 mark]